A Family Farm
in Tuscany

A Family Farm in Tuscany

Recipes and Stories from Fattoria Poggio Alloro

Sarah Fioroni

Translated by Natalie Danford

Shearer Publishing

Fredericksburg, Texas

Shearer Publishing
406 Post Oak Road
Fredericksburg, Texas 78624
Toll-free: 800-458-3808
Fax: 830-997-9752

www.shearerpub.com

The Library of Congress Control Number: 2011051201

Soft cover: ISBN 9780940672833
Hard cover: ISBN 9780940672840

Map illustration by Gloria Bani
Edited by Alison Tartt
Book design by Barbara Jezek

The photo of the author on the back flap is used with
the permission of *Organic Gardening* magazine.

Printed in Canada

*For my wonderful young
nephew and niece,
Ulisse and Vittoria*

*And for Umberto, Amico,
and Bernardo and their
beautiful families*

Contents

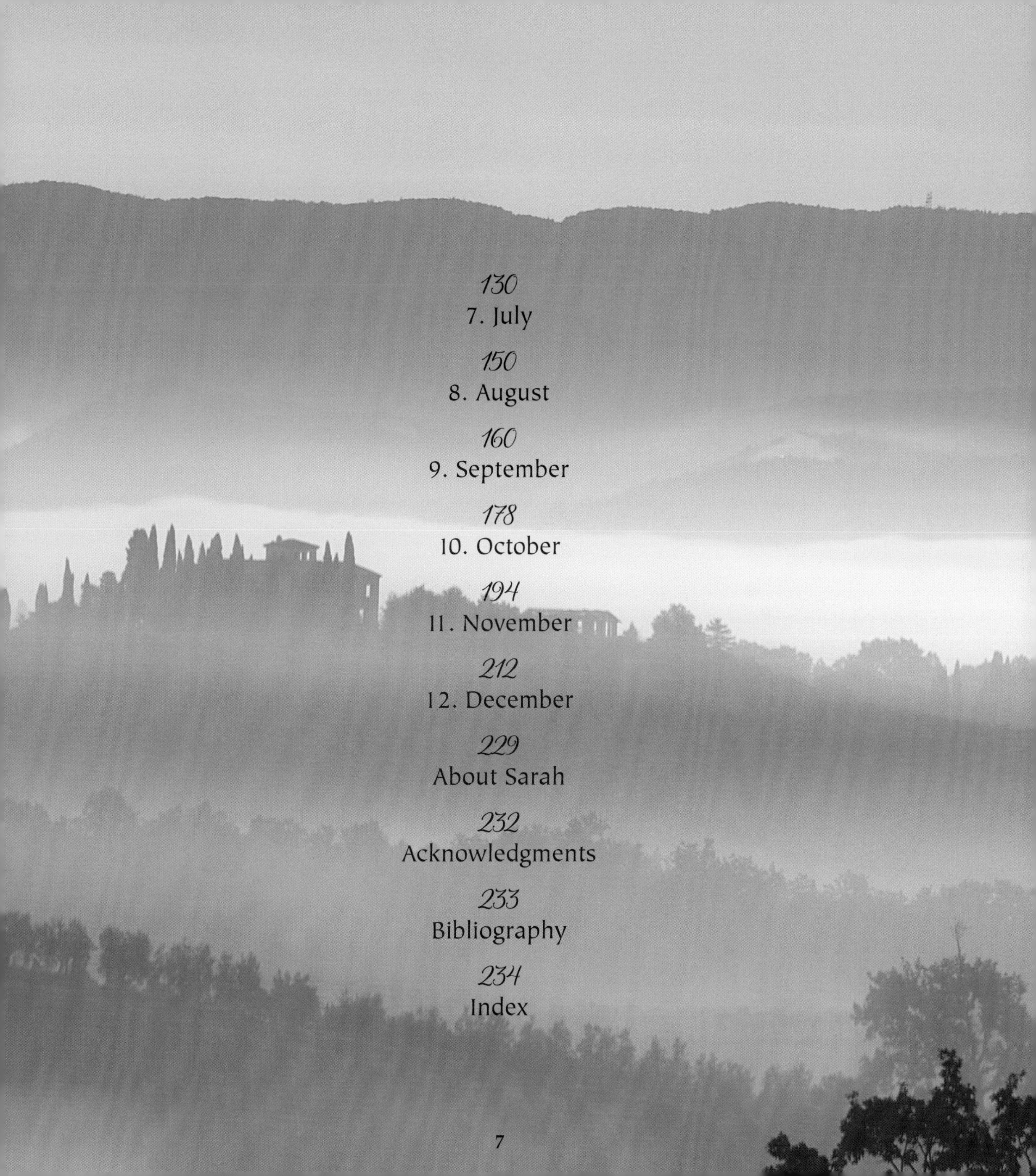

Preface

The concept for this book was born of my desire to share my family's story and our recipes with the largest number of people possible. Naturally, I couldn't go around telling our story to every passerby I saw, so I began writing these pages about what we do on a daily basis at the farm and things that happened to us in the past. This book is in no way intended to be a detailed account of typical Tuscan traditions; rather, it reflects the real-life experiences, both on the farm and at the table, of a single family.

This book is also a gift from me to my parents and to all the members of the Fioroni family. It cannot begin to repay people who spent their lives bent over the earth, working and making unbelievable sacrifices without ever asking for anything in return.

They did this for us, their children.

In doing so, they gave us the greatest gift that parents can give their children: a home, a livelihood, and a future. Most important, they taught us love for the land where we live.

While writing this book I repeatedly interviewed Rosa and Amico, my mother and father. Most of the time when they talked about the past, they had big smiles on their faces as they recalled those happy times when everything was simple and genuine; but occasionally they grew sad and even teary as they recalled difficult experiences involving poverty, war, and unfair treatment, things that are hard for us to understand and even harder for us to imagine.

Since I began teaching cooking, many people have told me that I should write a cookbook. At first, that didn't make a lot of sense to me. In Italy everyone knows how to cook, and very

few cookbooks are published. When I visited bookstores in the United States, though, I discovered a whole new world. It was then that I began to understand what writing a cookbook would entail and how wonderful it would be to share my family's story and excellent recipes with a wider audience.

Happy reading!

OUR STORY
The Fioroni Family and Poggio Alloro

The descendants of Amico Fioroni, my great-great-grandfather, were family farmers who owned some land and other property in the Marche region around 1800. Amico Fioroni had four brothers. With the death of their father, who had always handled the family's money matters, a brother who had planned to be a priest abandoned that idea and instead decided to take care of the family finances. He was the only family member who had gone to school, so his brothers handed everything over to him. He was not quite the businessman his father had been, however, and in just a few years he had frittered away all their assets and they were in debt to the bank. The family was in so much trouble due to these bad decisions that for several generations the members of the Fioroni family would be tenant farmers.

Life in the Marche

Sigilfredo Fioroni, the son of Amico Fioroni (1867–1917) and Palmina Michetti (1870–1953), was born in February 1897 in Montalto Marche in the province of Ascoli Piceno in the Marche region. He was the eldest of ten children, but five of them would die at a young age—in the war or of other causes. Sigilfredo was a farmer all his life, except for the period when he served in World War I with the Alpine regiment. The Alpine soldiers fought to free Italy from the German occupation of the Alpine region. Sigilfredo was captured twice by the Austrians and imprisoned in a concentration camp. He escaped both times. The first time he went into hiding in the

Opposite, left to right: Umberto, Amico, and Bernardo Fioroni.

FIORONI FAMILY

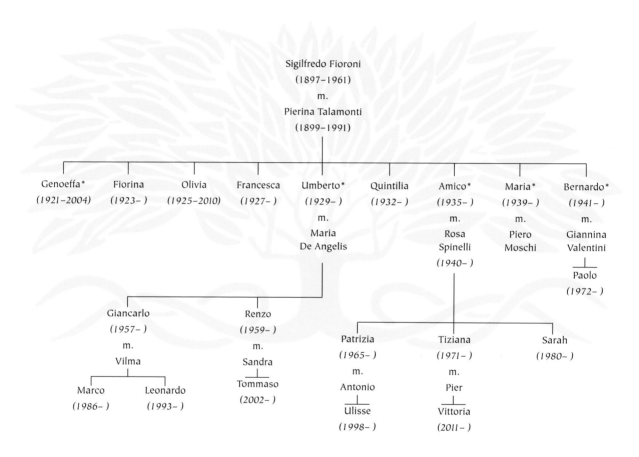

Sigilfredo Fioroni
(1897–1961)
m.
Pierina Talamonti
(1899–1991)

Genoeffa*
(1921–2004)

Fiorina
(1923–)

Olivia
(1925–2010)

Francesca
(1927–)

Umberto*
(1929–)
m.
Maria
De Angelis

Quintilia
(1932–)

Amico*
(1935–)
m.
Rosa
Spinelli
(1940–)

Maria*
(1939–)
m.
Piero
Moschi

Bernardo*
(1941–)
m.
Giannina
Valentini
|
Paolo
(1972–)

Giancarlo
(1957–)
m.
Vilma
|
Marco   Leonardo
(1986–)   (1993–)

Renzo
(1959–)
m.
Sandra
|
Tommaso
(2002–)

Patrizia
(1965–)
m.
Antonio
|
Ulisse
(1998–)

Tiziana
(1971–)
m.
Pier
|
Vittoria
(2011–)

Sarah
(1980–)

The siblings with an asterisk (*) are the ones who live (or lived) at Poggio Alloro or spend all their time there. The other living children of Sigilfredo and Pierina are married with children and live in the Marche and in Liguria. We still have a close relationship with them and see each other often.

home of a family in Udine. The second time he escaped near the woods along the Swiss border and hid there. Both of these escapes ended badly, however, and he was again captured and forced to work in a cement factory as a prisoner. When the war ended and he was free to return home, the twenty-year-old took up his life again as a tenant farmer. He met a neighbor, Pierina Talamonti (born in 1899 in Cossignano in the same province), and they married in 1920. The couple had nine children.

One of the youngest was my father, Amico Fioroni, who was born in 1935 in the small town of Petritoli, also in the province of Ascoli Piceno. The family was quite poor, and over time the living conditions under the sharecropping system in the Marche became intolerable. My grandfather even worked on a farm that raised silkworms, a backbreaking job. In 1944, as the Germans retreated, Germans and Italian Fascists came to the area to destroy everything. Sigilfredo and most of his sons managed to hide in the wheat fields. Only Pierina and Amico remained out in the open to show that the house was occupied. The Germans arrived and put a gun to my grandmother Pierina's head and stole two cows from her. They also took an older man as a hostage, then freed him four days later. Amico was terrified. They were lucky in that they survived and had managed to hide two other cows.

A New Start in Tuscany

After the war a lot of people began migrating from the Marche region to the region of Tuscany. By the 1950s, several families from the Marche had already moved to the San Gimignano area and spoke highly of it, saying that it offered many opportunities for work. Umberto, the oldest of my father's brothers, was the first to make the trip. He stayed for only a couple of days, then returned to tell his brothers what he had seen and what arrangements he had made in Tuscany.

Umberto says, "I arrived in Tuscany on the morning of December 27, 1954, and set off on foot to walk the six kilometers from Certaldo to the home of the Cameli family in the village of Santa Maria. They were the people who had written us a letter about Poggio Alloro. I saw the place and liked it, so I went to the owner, Signora Lida Landi in Certaldo. I signed a contract that said I would be charged a penalty of 50,000 lire if I didn't show up." He then returned home, and the family ended their sharecropping contract in the Marche.

Amico was the first to work on the farm; he came to Poggio Alloro in August of 1955 and worked briefly as a sharecropper. He was twenty at the time and had already performed eighteen months of compulsory military service in Sardinia. (Eventually compulsory military service in Italy would be shortened to twelve months, and it has now been abolished completely.) He then returned to the Marche for Umberto's wedding and to help with preparations for the move.

On October 23, 1955, Umberto arrived in Tuscany with his wife, Maria De Angelis, and his brother Bernardo. He says they were brought there by truck, like animals. Pierina came the next day, along with my aunt Maria as well as Francesca's husband.

Amico and Sigilfredo were the last to arrive. When they got off the train in Certaldo, it was evening and the buses were no longer running. So they called the only taxi service in the area—Fausto Maroni—to drive them to Poggio Alloro. It was probably dinnertime, but Fausto was no doubt happy to go pick up some fellow Marche natives at the station and take them to the farm. Coincidentally, Fausto Maroni also drove me to school, as he was our school bus driver, and he later started his own bus company, which now ferries passengers to Fattoria Poggio Alloro.

The Bandini family was already established in the area, and for many years the two families lived together on the farm. Poggio Alloro was an old dairy farm about five kilometers outside of San Gimignano that was surrounded by about twenty hectares of "good" land, meaning land that was suitable for farming. The farm is in a lovely spot, perched atop a small hill, with a spectacular view of the beautiful medieval towers of San Gimignano. As soon as my father saw it that summer, he sensed it was a special place.

At the time, my father was engaged to my mother, Rosa Spinelli, but she stayed behind in San Benedetto del Tronto in the Marche. Whenever I ask my mother about the first time they met, she tells me that she asked my father what Tuscany was like, and my father, trying to be charming and win her over, joked that the hens in Tuscany had brakes, meaning that Tuscany was very hilly. My parents married in 1964, and then my mother, too, moved to Tuscany.

Unfortunately, in 1961, when Sigilfredo was only sixty-four, he developed a blood clot in his brain. He lived only eight more days and then died, leaving Pierina the sole head of the household. Pierina rolled up her sleeves, and the nine siblings did the same. They hadn't gone beyond elementary school—secondary school wasn't compulsory in those days—but they were hard workers. She lived to see them grow up and buy the Poggio Alloro farm for themselves. She lived to see all her grandchildren and children get settled and lead productive, happy lives, and she herself lived to be ninety-two, always sharing a home with her family.

When I think of my grandmother Pierina, I remember her wrinkled face and hands, her kindness, and her thin, frail body. By the time I knew her, she was no longer the young, strong woman she had once been, but I was lucky enough to be held in her arms and to look into her beautiful blue eyes.

The owner of the land then was Lida Landi, born in 1879, who lived in Certaldo and owned several different properties, some of them just empty plots of land and others with buildings on them. She allowed the overseer to run the farm and dropped by from time to time to check on the work. The three brothers began to work as sharecroppers with little to their names. The sharecropping system, known as *mezzadria*, was basically a form of slavery that dated back to the Middle Ages. Sharecroppers farmed the owner's land in return for food and lodging, but at extremely low pay. In 1966 Italian law abolished the system completely.

The Mezzadria *System*

A *mezzadria* was a kind of farming contract that was regulated by law and that allowed the owner of a farm (both land and farmhouse) to assign work to other farmers, dividing the harvest and the profits from the land with them. However, the law was strongly tilted in favor of the landowner (*concedente*), who got 60 percent, while the farmer (*mezzadro*) got 40 percent.

This kind of contract dated all the way back to the Middle Ages and was based on the feudal system in place at that time. The overseer (*fattore*) was the landowner's representative; he took care of administering the farm and dividing up the harvest, and he wrote down everything in his accounting books. For his part, the owner recorded any animals that the *mezzadro* sold and took all the money. On January 31 of each year, the farm's totals were tallied. If there was a profit for the year, the *mezzadro* earned a little money; if not, he owed money. Often landowners were sneaky and played around with the books, showing losses that hadn't actually occurred. That meant the farmers ended up with nothing but had no way to prove that they were being cheated, as almost all of them were illiterate.

The farmer had numerous fixed obligations to the landowner, including the following "gifts" that he had to supply as tribute over the course of a single year:

Each month, the owner was to be given twenty eggs.

Each week, the women did the landowner's laundry at his house, or they took his laundry to their house, washed it, and then brought it back to him.

Before World War II, each man had to dig five kilometers of trenches for planting grapevines. This labor was unpaid.

Six capons, weighing eight to nine pounds (4 kg) each, had to be delivered to the owner at Christmas. If they were underweight, a debt was entered in the account books. In addition, six hens were required at Carnival, six chickens in August.

The owner was provided a pig that weighed 450 pounds (200 kg). If it weighed any less, the difference was entered as a debt in the accounting books, at a price chosen at the landowner's discretion.

The first fruit harvest had to be given to the landowner. (Amico says, "When I got a little smarter, I'd climb up into the cherry trees like a cat and really empty them out, thinking, better that I eat them myself than that they go to feed the boss!")

Chick peas, potatoes, and all the vegetables produced were divided up. Potatoes had to be dug up with either the overseer or the landowner present.

The system was not only unfair to people but bad for the soil because it encouraged overplanting and, as a result, overproduction of grapes, which meant they weren't of good quality. But of course people tried to produce as much as possible so there would be enough to eat once it was divided between the landowner and the farmer. The farmer's children were not allowed to go to school so they could contribute labor.

The farmer got a plot of land and an empty house, without heat, and with a little wood for the stove. At Poggio Alloro they had to forage in the woods for bushes or branches to light a fire. They certainly couldn't cut down any trees, or at least they couldn't let the owner know they were cutting down any trees. Every landowner had his "spies" who reported everything.

Many landowners were cruel and domineering. Amico remembers the time when his father, well into adulthood, received a slap across the face because he had forgotten to take off his hat when he bowed at the waist to greet the landowner.

In Tuscany conditions were a little better than they were in the Marche, where the plots of land were small and farmers were incredibly poor. After each harvest they found themselves left with practically nothing—it all went to the landowner. But in Tuscany there were large swaths of land and abundant harvests. Furthermore, the rules in this region had never been as harsh as they were in the Marche. (Farmers in Tuscany had to hand over two hocks of prosciutto to the landowner, while farmers in the Marche had to give up an entire pig.) And with the beginning of World War II, there were uprisings among farmers—they were organizing and making demands for agrarian reform.

By the time my family arrived in Tuscany, farmers no longer had to provide prosciutto

Clockwise, from upper left: Ladies of the family washing clothes. Pierina Fioroni. Maria Bandini and Maria Fioroni. Bruno Bandini (with dog). Piero Bandini (in white shirt), Maria and Umberto, Leonetto Bandini, baby Giancarlo, and Pierina Fioroni.

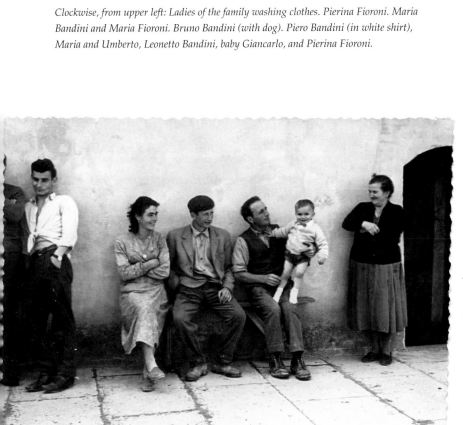

Top row, from left: Novaro Bandini. Leonetto Bandini.
Above left: Amico during the wheat harvest.
Above right: Umberto and Maria working with the harvested wheat.
Left: Sigilfredo and Pierina with baby Giancarlo.
Right: Umberto with a Chianina.

hocks to owners. Gradually, the *mezzadria* system was phased out. Beginning in 1964 new sharecropping contracts were forbidden by Italian law, although existing contracts were allowed to remain in force. A law enacted in 1982 provided for existing contracts to be converted into rental agreements. Yet in our area people were still working under sharecropping contracts until the 1990s.

Each year the Fioronis managed to earn 100,000 to 200,000 lire, a small fortune at the time. All the members of the family were animal lovers and did an excellent job of raising livestock, so they came out on the plus side. Equally important, the landowner, Signora Landi, was honest with them. She kept careful records and paid on time. She grew fond of my family very quickly, not least of all because she realized that they were tireless and honest workers.

The Purchase of Poggio Alloro

In 1964, the same year my parents married, the Bandini family moved to Poggibonsi, a town east of San Gimignano, so only the Fioroni family remained at Poggio Alloro. After that, only half of the land was used by the Landi family's workers because, with the Bandini family gone, it was hard to find new people to work the land.

The three Fioroni brothers were already working so hard that in 1972 Signora Landi, who had no heirs, decided to talk to my father about the fate of the land. She told him that her only concern was that she wanted to sell the land to people she could trust. She had decided that she couldn't possibly see it go to anyone other than the three brothers, because she had seen how they farmed the land with love and dedication, even though it didn't belong to them. She was sure they would know how to continue to manage and farm Poggio Alloro.

Umberto tells me that after the sharecropping system in Italy ended, a tenant farmer had a preemptive right to any land that the landowner put up for sale. The three brothers went to the Farm Inspectorate—an office that doesn't exist anymore—and asked about the land. They were told that there was a discounted 2 percent interest rate on a state loan of 33,000,000 lire. Signora Landi wanted to sell the property, but she wasn't sure that our family could shoulder the debt. In order to convince her, the brothers called the director of the inspectorate and asked for his help. He took the bus to San Gimignano, and Amico went to pick him up in a Fiat 500, the only car the family owned. Then he spoke with Signora Landi, who was ninety-four by then but still very sharp mentally, and explained the state farm loan system to her. Finally, she was convinced, and they negotiated a price of 45,000,000 lire for Poggio Alloro. Today that would equal about 23,000 euros, or the price of a car!

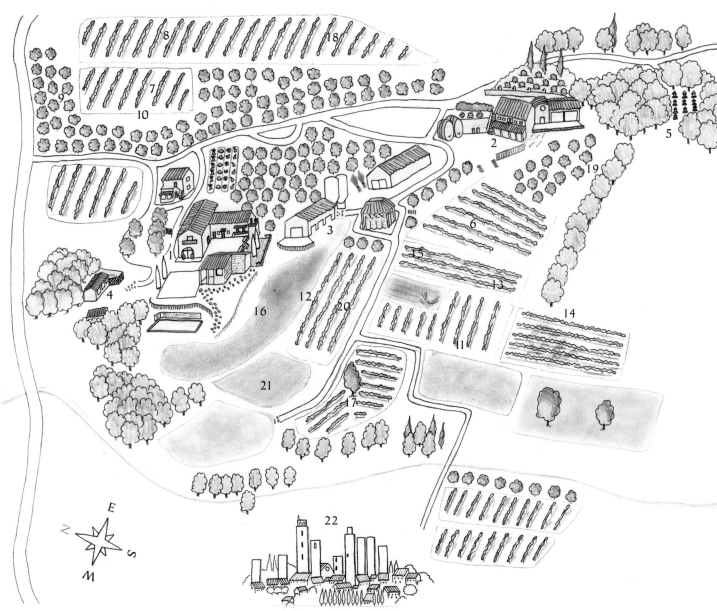

1. Fattoria Poggio Alloro – office and restaurant
2. Cantina – Wine Cellar
3. Stalla – Barn
4. Animali de barra corte – Barn
5. Arnie – Beehives
6. Campo del Sorbo – Rowan Tree Field
7. Il Nicchiaio, Poggio di Bista
8. Buca della Fonte – Fountain Hollow
9. Campo degli Olivi – Olive Field
10. Pozzo degli Zingari – Gypsy Pond
11. La Fonte – The Spring

12. Meli Schiaccioli
13. Campo dei Ciliegi – Cherry Tree Field
14. Poggio delle Tortore – Turtledove Hill
15. Piana delle Grillaie – Grillaie Plain
16. La Buca del Dinosauro – Dinosaur Hollow
17. Poggio del Castagno – Chestnut Hill
18. Poggio dei Mandorli – Almond Hill
19. Ginestraio
20. Fosse Lunghe – Long Trenches
21. Il Lago – The Lake
22. San Gimignano

So on February 23, 1973, the three brothers agreed to purchase the property from Signora Landi. They entered into a thirty-year mortgage, and Lida Landi lived to be 103.

In those days, a farmhouse consisted of bedrooms and a kitchen, with a fireplace upstairs and the stalls for the animals on the lower floor. Often the bathroom was outside the house. Once they had purchased the farm, the brothers began to rebuild the farmhouse so that each family would have its own apartment. It would be their first time living in separate quarters.

In 1980 they also built a cowshed at the foot of the hill for the Chianina cattle they were raising, which put the animals farther away from the house. Until that time, the animals had been kept on the first floor, in part so that they could heat the second floor. In 1983, down in the valley, they dug an artificial lake that would be used to water the fields and also to provide the animals with drinking water. Each year they planted additional grapevines and olive trees.

About 400 years ago, the property we now call Poggio Alloro was known as Il Venzuolo. Historical records show that our farmland actually belonged to the family of Fina dei Ciardi, otherwise known as Saint Fina, the patron saint of the city of San Gimignano. Centuries ago, land was simply given a name—there were no land registers with plots neatly measured out. Properties were defined by their physical features and natural boundaries.

The name Poggio Alloro means Bay Leaf Hill, and it refers to the bay laurel bushes that grow everywhere on our property (a *poggio* is a small hill). The land was planted with olive trees like those we still cultivate today.

There was also Campo del Sorbo, or Rowan Tree Field, with large trees and rows of field maples, trees that grew wild and that don't exist anymore. The piece of land that now stands next to the province road was called Il Nicchiaio, or just Poggio di Bista. (Bista was the name of a worker employed by Signora Landi.)

Buca della Fonte (Fountain Hollow) was a piece of land at the bottom of the valley where there was a freshwater spring. Animals were taken there so they could use it as a watering hole. The area was filled with wild poplars.

The Campo degli Olivi, or Olive Field, was covered with centuries-old olive trees that were killed in the 1985 freeze. This land faces east, and toward the top is a little area of woods with a pond known as the Pozzo degli Zingari (Gypsy Pond). This pond provided water in the summer, and it looked like a little lake surrounded by woods. After we hooked up a system to get our water from La Fonte (The Spring), another pond in the valley, all the young people would go swimming in the Pozzo degli Zingari.

Today there is a new vineyard in the areas once known as Meli Schiaccioli and Campo dei Ciliegi. These two plots of land take their names from the trees that grew on them—apple trees (*schiacciola*, a variety that is hard to find today) and cherry trees, respectively.

Poggio delle Tortore (Turtledove Hill) got its name from the fact that hunters often found turtledoves there during hunting season. Beyond this hill is the Piana delle Grillaie (Grillaie Plain), a piece of land where no one was able to get anything to grow, so it was left to the chirping crickets for which it was named.

The land where our Chianina cows graze today was known as La Buca del Dinosauro (Dinosaur Hollow). Hundreds of years ago some vertebrae that didn't belong to any farm animals were found there.

Poggio del Castagno (Chestnut Hill) retains its name today, even though that hill was treeless or simply seeded for many years. Today it is the site of a beautiful vernaccia vineyard, and our only remaining wild chestnut tree stands there as well.

Poggio dei Mandorli (Almond Hill) is now called Vigna Le Mandorle, or Almond Vineyard. Our highly regarded vernaccia wine is labeled Vernaccia "Le Mandorle," both as a nod to this hill and to the fact that vernaccia often has an aftertaste of bitter almond. Our best vernaccia grapes come from this area.

The four wooded hectares to the southern side of the property were known as the Ginestraio, named for the gorse shrub that grew there.

Finally, Fosse Lunghe, or Long Trenches, got its name from the trenches that were dug by hand to plant rows of grapevines and field maples, which were planted together using a system called *alberello* ("little tree") or *vite maritata* ("married vine") because the vines were "married" to the maples in the sense that the maple branches were used to support the vines as they grew. We no longer use this "romantic" system, instead relying on cement posts and wire.

Poggio Alloro Today

Fattoria Poggio Alloro consists of one hundred hectares of land used for a variety of purposes. The farm (*fattoria*) is completely organic and has been for fifteen years. For the twenty years before that, we were 80 percent organic. Going organic was not a political statement, nor was it a strategy to turn a profit. Quite simply, in the 1950s the only farming methods my family knew were organic methods. Organic farming really is in our blood.

On twenty hectares we grow grapes typical of the area, such as vernaccia di San Gimignano, sangiovese, canaiolo, colorino, red and green malvasia, trebbiano, cabernet, merlot, chardonnay, and San Colombano. We use the grapes to create wonderful, high-quality wine. Our grapevines don't need any irrigation as they get enough water from rainfall.

On another twenty hectares we raise grains such as durum wheat and farro, which are processed at a mill to make delicious durum semolina and farro pasta. Other crops like barley, oats, corn, and sunflowers are used to feed the animals or sold as grains.

We make excellent extra-virgin olive oil with the fruit of 1,500 olive trees, which grow everywhere on our property, even around the house. Our garden and orchards also yield numerous types of fruits and vegetables.

In winter we rely on wood from the farm's two wooded areas. One of these is also the site of thirty-five beehives, where we keep the bees that produce precious nectar in the form of millefiori honey.

We still raise Chianina cattle, just as we always have. The Chianina breed is now at risk of extinction. We use a closed breeding program with the Chianina cattle, meaning that we never buy outside cattle. They're born and butchered on the farm, and all of their feed is completely organic. We do use artificial insemination to help them reproduce, however, as keeping a bull is very costly, and with only twenty-five breeding cows on average, we could easily inadvertently breed among members of the same family.

Our courtyard animals—chickens, rabbits, and pigs—live in wide-open spaces and graze freely.

Tuscany was one of the first regions in Italy to experience the rise of *agriturismo* (farm tourism), which gives tourists opportunities to participate in farm life, learn about farming operations, and taste a farm's products. In 1991 a regional decree was passed that established a program for renovating preexisting farm structures as lodging for tourists. This program offered some assistance to farmers at a time when agriculture itself was not yielding great results.

Through this program we created six double rooms for tourists by renovating one of the buildings that had previously been used to house animals. In 2000 another small building was renovated to create an additional four double rooms. At that time we also dug a beautiful swimming pool with a view of San Gimignano.

Guests at Poggio Alloro are given the chance to be surrounded by nature and animals and to taste local dishes prepared with the farm's products. They go for walks and bike rides, work on the farm, and participate in tastings, guided tours of the farm, cooking classes, and other types of educational experiences.

Best of all, guests wake up in the morning and gaze out at the incredible landscape and enjoy the relaxing serenity of this magical place. Many of our guests call it paradise! Mornings are sweet, with a breakfast that consists of fresh milk, provided earlier that morning by our own cows, and baked goods still warm from the oven.

Every day we serve what we call "light lunch"—although many guests complain that it's a lot of food! We serve the lunch, accompanied by wines, on our beautiful terrace with its amazing view of San Gimignano. A typical lunch consists of bruschetta and pasta, a fresh salad dressed simply with extra-virgin olive oil and salt, pecorino cheese, prosciutto and salami, ending with a glass of *vin santo* and *cantuccini* cookies.

Dinners are served at a long table where our guests coming from different countries and speaking different languages all become friends after a good glass of wine! The four-course dinner menu starts with a large selection of antipasti (appetizers), pasta, or sometimes soup, then meat with side dishes, and finally a dessert, plus *cantuccini* and *vin santo,* coffee, grappa, and *limoncello.*

We also operate a little store all year long where all the farm products are available as well as our eight types of wine, *vin santo,* several sizes of extra-virgin olive oil, pasta, cookies, honey, saffron, grappa, and, in the winter months, fresh beef and pork. We even have nice souvenirs—such as farm aprons, hats, and t-shirts. Everyone always wants to buy "a piece of the farm to take back home!" as some of my guests like to say.

From the veranda a beautiful panorama of rolling hills and lush fields stretches as far as the eye can see, with the towers of San Gimignano nestled on the horizon. Each day the weather conditions subtly change the look of the towers—on some days they appear far away and tiny, while on other days they are so clearly visible and bright that you feel as though you could almost reach out and touch them.

The question I'm asked most often is whether I ever get tired of living in the same place all the time.

How could you ever get tired of paradise?

Opposite, left to right: Giancarlo, Paolo, Renzo, Sarah, and Marco.

A BRIEF HISTORY OF SAN GIMIGNANO

S an Gimignano is a small but very special city located in the heart of Tuscany, thirty kilometers from Siena and forty kilometers from Florence. San Gimignano sits atop a hill, a little over 1,000 feet (324 m) above sea level. It is known for its beautiful towers, which make it recognizable from a distance. Today San Gimignano has a population of about 7,000—the same number of residents that it had in 1227 and in 1850.

The city developed quite rapidly in the Middle Ages, but it dates back even farther than that. The first settlements in the area were created as early as the sixth century B.C., when Etruria was thriving. Etruria was an ancient region located in what is now central Italy. It included the southernmost part of Liguria, Tuscany, and parts of the Umbria and Latium regions. Archeological research tells us that there were residences, burial grounds, and rural properties in the area at that time.

That all came to an end when the Roman Empire crumbled and the Barbarians and Byzantines invaded. There are no traces of Etruscan civilization after that time.

Around the seventh century A.D., the first roads were built in the north, and the Via Francigena, which ran from Frankish land to the center of Christianity, was constructed. This road linked northern and southern Europe, all the way from Canterbury, England, to Rome.

In 1350 Pope Boniface VII established the first Jubilee, and Rome became a pilgrimage site for the forgiveness of sins. Via Francigena also provided indirect access to two other pilgrimage sites: Santiago de Compostela and Jerusalem. The areas through which this road passed all developed quickly and became important commercial sites. Construction of the road led in turn to construction of inns, taverns, customs houses, sanctuaries, pilgrim hospices, and markets. Via Francigena ran right through San Gimignano, and soon the city was a very active place.

During the same period, a large castle was built on the city site. In the ninth and tenth centuries, the castle began to have its own internal structure. At the same time, the more removed historic central area was built and then walled off. The walls both kept the population in the town and served as defense. Construction of San Gimignano's now famous towers began in the tenth and eleventh centuries. Often the base of a tower was used as a workshop or store, while the upper floors were residences. Towers were often equipped with small wooden balconies that faced the street. Around 1200 this type of building came to a halt. The only tower dated after that period is the Torre Grossa, literally the "big tower," built in 1300.

During the era of the city-states, these centers demanded increasing power as they sought independence, to the detriment of the church's power. A *podestà* was named to maintain public order and govern the city-state. The podesta was an outsider, usually knowledgeable about legal matters. He represented the head of the army and oversaw the surrounding countryside, any production, and the defense of local resources.

In 1200 the population was expanding, and as a result another wall was built. Around the middle of that century, the city was divided into four *contrade*, or neighborhoods: Piazza, Castello, San Giovanni, and San Matteo.

This was a period of growing internal tension, as the land in the area became more valuable. It took very little to ignite conflict between the old landowners and a new class of emerging families who were involved in artisanal production and organized into guilds.

Throughout Italy in this period, people divided into two groups: the Guelphs, who supported the pope, and the Ghibellines, who supported the emperor. San Gimignano was not untouched by this rift, and often family members were divided by their loyalties. There was a fierce rivalry between the Salvucci and Ardinghelli families, which led these two factions. The situation calmed down in late 1200 as a series of alliances was formed. Even as they struggled against each other, both factions understood the need for stability inside the castle walls. This was a glorious time in San Gimignano's history, with much of the population enjoying great wealth and a comfortable lifestyle.

In 1273 the well (*cisterna*) was built that would give the Piazza della Cisterna its name. At the time, every palace had a well or water tank to supply its own water. In 1298 the Palazzo del Podestà, which would become the center of secular power, was completed.

On May 8, 1300, Dante Alighieri, author of *The Divine Comedy,* arrived in San Gimignano. As a prior of Florence, he came with the intention of recruiting members to the Guelphs. He spoke in a room that would thereafter be known as the Dante Room. It was also during this period that painters created the magnificent frescoes in the Duomo that can still be seen today.

From the late 1200s to the early 1300s, Via Francigena, which had brought such fortune to the city, was changed slightly. Its new route ran toward the valley in the Elsa River area, which was drained. Toll revenue along the road still continued to be high, thanks to its strategic position.

The region's economy was based on agriculture. The area was very fertile. Food for animals was grown there and sold and purchased at a market that remained active until the 1400s. Vernaccia, olive oil, and saffron were all produced here at that time.

The people of San Gimignano were also involved in money lending. The bishops of Volterra were in debt, and they took loans from the Florentines as well as the people of Siena and San Gimignano, who charged very high interest rates. The local farmers were at a great disadvantage under this system. They were often unfairly manipulated, and when they were unable to make loan payments, the lenders would confiscate their harvests and resell the items at a high price at market.

In the early 1200s the city was also an active spot for merchants, who bought goods from the major outlets and resold them on various markets. Because San Gimignano had close ties with Pisa, it had access to the sea. Soon, however, San Gimignano found itself involved in a dispute between Florence and Pisa. The city asked Florence to cease its hostilities with Pisa so that local commerce would not suffer. But Florence refused and forced San Gimignano to choose between the two. San Gimignano broke off relations with Pisa and lost access to wool, fabric, leather, cheese, wheat, and spices. Production continued to be high, but goods produced in San Gimignano could then be sold only on the local market.

By the late 1200s, the city was closely associated with Florence, which controlled the area due to the appointment of Florentine podestas. As for the city itself, it was controlled by the bourgeois class and no longer in the hands of the aristocracy. In general, it was these bourgeois families who built San Gimignano's towers and created the city as we know it today. By 1355 there existed a total of seventy-two towers. They symbolized the wealth, power, and prestige of the families that owned them. Today thirteen towers remain.

In 1358 Florence ordered San Gimignano to build a fortress at its own expense. The city was forced to take loans from the Florentines to do so, which tightened Florence's control. In the fifteenth century, the era of the towers came to an end. They were neglected and began to fall into ruin. Pieces that fell off of the towers were often used to build other structures. The look of the city center changed, as did the facades of buildings. An attempt was made to remodel the towers that remained standing to make them look less medieval.

Farming had shifted to the *mezzadria* sharecropping system by this point, which meant that many of the fields were abandoned. Artisanal work, too, was in decline.

In the mid-1500s, under the reign of the Medici family, San Gimignano recovered and again became an important center for commerce. When the war with Siena ended, military structures were dismantled. The population again began to grow.

But then in 1630 a plague hit the city. Attempts were made to stop the plague by having inspections at the walls, but that was not enough. Quarantine was declared. The last death from the plague occurred in October 1631. In all, 1,000 people are thought to have died. Recovery from the plague was slow and difficult. Only around 1784 did the city truly begin to recover. San Gimignano then became part of the province of Siena, ending the love-hate relationship that had linked it to Florence for so many years.

Farming increased, which led to a bit of an economic recovery, as the resulting food could feed the local population, both farmers and those who lived in the city.

In the 1800s tourism became more common around the world: It was then that the first guidebooks were printed, and the idea of the "grand tour" gained traction. San Gimignano, which had remained virtually unchanged since medieval times physically, became a popular tourist destination. Maintenance and renovation of the city's buildings began in earnest.

In 1944, during World War II, the city was bombed. Attentive visitors can still spot the damage to the walls, the buildings, and the towers.

Today San Gimignano has returned to its former glory. It is a protected UNESCO World Heritage Site, and the area is subject to strict regulation that has kept overbuilding at bay. As a result, a visitor arriving in San Gimignano now has an opportunity to experience the feeling of going back in

time to the Middle Ages. Enter San Gimignano through one of two gates, Porta San Matteo or Porta San Giovanni, and stroll through the town. On the streets there are many stores selling handcrafted products, places to stop for a drink or a bite to eat, and the famous *gelateria* in Piazza della Cisterna, which sells the world's best gelato! The Palazzo Comunale, the Torre Grossa, the Duomo, and other churches can all be visited.

The city boasts numerous excellent museums with unique, prestigious collections. For further insight into San Gimignano as it was in 1300, just visit the San Gimignano 1300 Museum, which houses a miniature model of the city as it existed during the Middle Ages. Rocca di San Gimignano, an old fortress, now hosts a lovely green park area and the Museo del Vino Vernaccia di San Gimignano.

Wine's cultural and economic importance is highlighted by many local events and attractions, such as those sponsored by the Associazione Strada del Vino Vernaccia di San Gimignano. Formed in 1999, this is one of several Strada del Vino (Wine Route) associations throughout Italy that promote wine and food appreciation and offer tourists not only wine-tasting itineraries but also general information about the area.

San Gimignano's Strada del Vino association, for example, provides information on wineries and farms to visit, which ones offer wine and food tastings, and where visitors can buy wines and other typical local products. The association also organizes various shows during the year that are open to tourists and local residents alike. These include *pomeriggi dei produttori,* or winemaker afternoons. At these events, held in the beautiful old San Gimignano fortress, participants have an opportunity to taste wines from three winemakers as well as a local dish served by a restaurant every Friday from June to September. The association also runs something called *degusta con noi* (meaning "taste with us"), which consists of visits to wineries with guided tastings twice a week. What's more, during the summer there are the monthly *serate a tema,* or theme nights. Attendees have a chance to taste products from various San Gimignano producers, and area organizations provide entertainment by playing music or performing typical local folk arts.

One memorable performance was provided by the theater company Comici Ritrovati. Members recited prose and poetry dating from the Middle Ages to the present day, all having to do with wine. The actors surprised everyone by mixing in with the crowd of tourists and tasters and then reciting their lines with glasses of wine in hand. It was a truly special evening under the stars. Additionally, the association operates the wine museum in the Rocca, which sees thousands of visitors a year and is completely free.

The tenth anniversary celebration of the San Gimignano Strada del Vino association took place in the beautiful central piazza in San Gimignano, Piazza Sant'Agostino. Local winemakers and restaurateurs, the real stars of the show, not only provided wine and food free of charge, but also generously contributed the kind of time and hospitality required for such an event. Attendees could purchase glasses and glass holders for a small fee, and then they were free to tour the many tables arranged in a horseshoe around the piazza and taste excellent wine and foods. The featured entertainment was provided by two groups: the Cavalieri di Santa Fina from San Gimignano and the Sbandieratori dei Borghi e Sestieri Fiorentini from Figline Valdarno, whose members wore medieval costumes and performed a drum-and-flag routine.

Cavalieri di Santa Fina

This group of medieval reenactors was founded in 1993 and named for Fina dei Ciardi, who died on March 12, 1253, after having dedicated the final years of her tortured life to God. The day of her death, violets miraculously bloomed on the walls of San Gimignano, including the walls of its famed towers. She was made a saint and became the patron of the city, along with Saint Gimignano himself.

The Cavalieri di Santa Fina celebrates certain aspects of San Gimignano's courtly tradition. Every year, it stages the Ferie Messium, or Fiera delle Messi, a celebration of the harvest season that dates back to the Middle Ages. It consists of knights on horseback, drummers, archers, medieval dancers, the armed group Compagnia della Rocca di Montestaffoli, and reenactors. Working together, they put on a very evocative show.

The group is further divided into four *contrade*, representing the sectors of the historic center of the city. When they're around, it's really difficult to tell whether you're in the Middle Ages or the present day! The drum rolls that announce their arrival only serve to heighten expectations. When the group arrives in medieval formation, you can see all the traditional court figures, from drummers to knights to soldiers with ramrod-straight posture to beautiful ladies.

The procession marches to Rocca di Montestaffoli in the historic center, where they perform the ever-popular *giostra dei bastoni*, a joust in which the four *contrade* challenge each other. It may seem like a dream, but it's all very real!

Members of Cava-lieri di Santa Fini display elements of San Gimignano's traditions of medi-eval pageantry.

The Sbandieratori dei Borghi e Sestieri Fiorentini exhibit the military artistry of flag throwing for spectators in San Gimignano.

Sbandieratori dei Borghi e Sestieri Fiorentini

T his group, founded in Figline Valdarno (in the province of Florence) in 1965, consists of about eighty different members (all of them men) and is divided into captains, drummers, clarion players, and flag holders. The group's flag displays half of a fleur-de-lis and a lion, which is also the symbol of Figline Valdarno. The group is well known internationally and was even part of the opening ceremonies for the Frankfurt World Cup in 1974. Performances evoke the tradition of both military and performing arts.

In the Middle Ages the presence of flags among local troops was a sign of patriotic pride in one's city. Carrying flags was also a practice that existed for tactical reasons—during combat, a soldier could locate his squadron by means of the right flag. If a flag fell, it would be captured by the enemy.

Today's performances include various flag movements that hark back to those times, and the shows are very dramatic and elegant to watch. First the audience hears the sound of the drums. Then the flag holders appear in various formations, wearing white and blue and white and red costumes. An exhibition with flags ends with a victory celebration that consists of crossing the flags, a very precise, colorful flourish that keeps time with the rhythm of the drums and horns. The eyes of every audience member are glued to the flags as they are tossed high in the air, back and forth, with such skill that it looks like child's play. The whole show is truly a jaw-dropping spectacle.

JANUARY

Snow has fallen during the night. Everything is silent. There are no sounds of life—no cars, no animals, just the occasional footstep crunching in the snow. It's incredibly peaceful.

When we were little, the moment on snowy mornings when we'd open the windows and see white all around was a magical one. Sometimes there was an amazing amount of snow on the ground. We thought it was great fun to sled as fast as possible down the slopes of the snowy fields. Actually, our "sleds" were big plastic sacks that had originally contained salt. We'd sit on top of them, holding the ends in our hands, and then we'd sled down like crazy. Sometimes there were two or three of us on the same sled. By the end of the day, our rear ends were killing us, but

we were having so much fun that we never felt cold or tired.

On this snowy day I'll be eating polenta for lunch. Sitting all the way over in my office, I can smell the polenta getting underway. When I arrive in the kitchen, I find my aunt Giannina, who is tiny, energetically mixing polenta in a pot with a wooden stick. We use cornmeal made on the farm from our own corn. It has to be cooked for several hours until it's firm. It's a filling and nutritious dish that's perfect for warming up on a cold day like today. Polenta was very often served by farming families as lunch on its own, with no second course, as it cost very little and filled the stomachs of hard-working family members until dinnertime.

One by one, we take our seats at the

table. More and more we tend to eat all together in the dining room in the old house— a return to our old way of doing things. Everyone in the family who works with the *agriturismo* guests and the groups that come for tastings, all the waitstaff, and even some of the field hands eat together, so there are always ten to fifteen people around the table. Each time we get together, we tell stories about funny things that have happened at the farm over the years.

We sit around, laughing and relaxing, until 12:30, when it's time to have our coffee and get back to work. We still have a long day ahead of us.

In January the fields are at rest. There are still animals to be cared for, of course, but most of the work during this time of year takes place inside the house. My father often sits on a small stool by the fireplace and weaves branches into a wicker basket. The branches come from willow trees and are also used to tie up the vines in the vineyard. Many farms use synthetic ties, but we still use willow branches. Weaving the base of the basket is the most difficult part. Then Amico takes one branch at a time and weaves it into the base to form the sides of the basket. This meticulous and precious art is unfortunately now all but lost.

The biggest holiday in January is Epiphany, which falls on January 6. On that night, stockings full of sweets—or coal, garlic, and onions for bad little children—are hung above the fireplace. Italian folklore has it that La Befana, an old lady who rides from house to house on her broomstick, drops in through the chimney to leave the stockings for children.

Butchering the Pigs

Butchering a pig has always been a cause for celebration in our family. Pigs are butchered in the winter, as the work must be done in cold temperatures. It takes three days to butcher a pig. And no part of the pig is ever discarded!

On the first day, a pair of pigs are slaughtered. Their blood is drained and set aside. Their bristles are removed so they can be used to make brushes. Then the animals are split in two, and that's when the real work begins. First, they are chopped up. Each pig provides two shoulders and what will become two hocks of prosciutto. Women clean the guts *(budella)*, including the intestines and organs, which will be used as casings for salami. The flesh is left to rest and hang until the following day. On the second day, the parts are divided into smaller cuts, such as ribs and chops. Other meat is set aside to be ground for sausages and salami.

The third day is the most fun, the most celebratory, because it's the day we make the salami. There are various grinds of meat, some with a finer or coarser consistency than others. Then we add the spices and flavorings—like pepper, garlic, red wine, and so forth—according to my father's secret recipes.

We make sausages, Tuscan salami, fennel salami (with a finer grind and wild fennel seeds), *lonza* (also called *capocollo*), and pancetta. And since nothing is ever thrown away, even the head of the pig is boiled and chopped into pieces, then flavored with spices and lemon and orange peel and stuffed into a linen bag sewn by the women—it's called *sopressata*. Yet another type of salami, *buristio,* is made of pig's blood mixed with portions of the head and then stuffed into an intestine and boiled.

The best things, though, are *migliacci,* crepes made with fresh pig's blood, sugar, orange zest, *cavallucci* (a kind of Tuscan cookie), stale bread, and cinnamon. This mixture is poured into a pan, cooked like a crepe, and then served hot with sugar sprinkled on top. *Migliacci* are so delicious that we kids often ate them until our stomachs hurt.

During the last day of butchering, we would all eat together at a long table in the big room with the fireplace, the oldest room in the house. Lunch on that last day always

Clockwise, from top: Marco. Amico weaving a basket. Bernardo clearing snow.

Top, far right: Amico preparing proscuitto di cinta senese.
Above, left: Rosa making finocchiona, *or fennel salami.*
Above, right: Graham.
Far left: Maria Moschi.

Polenta Fritta
Fried Polenta

This dish can be used as an appetizer or a side dish and is really delicious.

Serves 10

6 cups (1.4 L) cold water

2 cups (300 g) instant polenta

1 cup (90 g) grated Emmental cheese

1½ tablespoons (21 g) unsalted butter

2 teaspoons (13 g) sea salt

2 egg yolks, beaten

2 cups (500 ml) canola oil for frying

BREADING

3 eggs, beaten with ½ teaspoon (3 g) sea salt

3 cups (450 g) dry breadcrumbs

Bring the water to a boil in a heavy-bottomed, 6-quart (6 L) saucepan over high heat. Slowly add the polenta, whisking continuously, until the polenta begins to thicken. Add the cheese, butter, and salt; whisk until the cheese is melted. Whisk in the egg yolks, incorporating them thoroughly. Cook an additional 4 minutes, whisking constantly. Remove from heat.

ʚɞ

Turn the polenta out into two prepared 9 × 5-inch (26 × 10 cm) loaf pans, filling each about two-thirds full and spreading evenly. Refrigerate until the polenta is cool and set, about 2 hours. When polenta is well chilled, turn out onto a cutting board and cut into slices about ⅜ inch (1 cm) thick. Set aside.

ʚɞ

Place the beaten, salted eggs in a wide bowl and the breadcrumbs in a baking dish. Dip each slice of polenta in the beaten eggs, coating both sides; then dredge in the breadcrumbs, coating well on both sides and shaking off excess crumbs. Set aside.

ʚɞ

Place a wire rack over a baking sheet and top with absorbent paper towels. Heat the canola oil in a heavy 12-inch (32 cm) skillet over medium heat to about 350 degrees F (175 degrees C). Fry the polenta slices in the hot oil for about 4 minutes, turning once, or until golden brown. Drain on the prepared rack and continue frying until all of the polenta slices are cooked. Serve about 3 slices per person.

consisted of a dish of polenta that the women began to prepare early in the morning in a pot that hung from a hook and chain in the fireplace. The polenta was served on a flat platter, and a sauce made with the pig's lungs and spleen was poured on top. It was hot and good.

That evening there would be a big party, and finally we could eat some pork! Dinner always began with a rich dish of pasta with leeks and sausage. For some people enough was never enough, so they would set a piece of leftover polenta over the fire on a grate to char it, topping it with extra-virgin olive oil and Parmesan.

Then the competition, especially among the younger cousins, really heated up. Many, many ribs were cooked over the fire, and there was always a contest to see who could eat the most. As the youngest, I gave it my best shot, but I never did win!

The evening ended with stories and jokes, and especially with teasing older relatives about the way they talked: they still spoke a dialect from the Marche.

Slowly people would begin to leave. Those who remained played card games or simply sat by the fireplace until the fire went out and their eyes stung from the smoke. Then and only then was it time for everyone to go to bed.

As we did many years ago, my family gathers in the oldest room in the house in winter to eat together during the pig-butchering period. We sit around the same table in the same seats, and the dishes still taste the same. It's always wonderful to gather together after a year of hard work.

Crostini con Mozzarella e Salsiccia

Crostini with Mozzarella and Sausage

Serves 6 to 8

1 French baguette, sliced ½ inch (12 mm) thick

1 pound (454 g) fresh mozzarella cheese, sliced ¼ inch (6 mm) thick

1 pound (454 g) Italian sausage, cooked and sliced into thin rounds

Preheat oven to 350 degrees F (175 degrees C). Place the baguette slices on a heavy baking sheet and bake in preheated oven for about 5 to 7 minutes, or until toasted.

☙

Remove from oven and top each crostini with a slice of the cheese and a couple of slices of the warm sausage. Return to oven and bake for about 8 minutes, or until the mozzarella is melted. Serve hot.

Opposite, upper left: Donatina serving tegamata.
Left: Bernardo.
Above, left to right: Sarah, Vilma, Giancarlo, and Renzo.
Right: Andrea, Maria, and Piero Moschi.

Crostini Colorati

Colored Crostini

Serves 8

1 French baguette, sliced ½ inch (12 mm) thick

5 ounces (140 g) low-moisture or processed mozzarella cheese, sliced thin

9 tablespoons (135 ml) extra-virgin olive oil

1 garlic clove, peeled and minced

6 white mushrooms, sliced

1 sprig Italian parsley, minced

Sea salt to taste

1 yellow bell pepper, cut into thin julienne slices

1 ripe tomato, cut into ¼-inch (6 mm) dice

2 uncooked Italian sausage links, about 4.5 ounces (135 g)

3 marinated artichoke hearts, quartered

Preheat oven to 350 degrees F (175 degrees C). Line a heavy-duty baking sheet with parchment paper. Arrange the baguette slices in a single layer on the baking sheet. Place a slice of cheese on each bread slice; set aside.

Heat 3 tablespoons (45 ml) of the olive oil in a heavy-bottomed, 10-inch (25 cm) skillet over medium heat. When the oil is hot, cook the garlic just until it begins to brown. Add the mushrooms and parsley. Season with salt. Cook over medium heat for 15 minutes, or until mushrooms are lightly browned. Turn out into a bowl and set aside.

Wipe out the skillet and return to heat, adding 3 tablespoons (45 ml) of the olive oil. When oil is hot, sauté the strips of yellow bell pepper until they are wilted, about 10 minutes. Season with salt, turn out into a bowl, and set aside.

Wipe out the skillet and return to heat, adding the remaining 3 tablespoons (45 ml) of olive oil. Remove the sausage from casings and crumble with your fingers. When oil is hot, cook the sausage until lightly browned, about 4 to 6 minutes. Remove from skillet with slotted spoon and set aside.

Top each bread slice with one of the following toppings: the mushroom mixture, the bell peppers, the diced tomato, the sausage, or the artichoke hearts. Bake in preheated oven for 8 minutes, or until the mozzarella is melted. Serve hot.

Pasta Salsiccia e Porri

Pasta with Leeks and Sausage

Serves 6

4 tablespoons (60 ml) extra-virgin olive oil

3 cups (265 g) leeks, sliced ¼ inch (6 mm) thick

⅛ teaspoon sea salt (1 g), plus additional for the cooking water

3 tablespoons (45 ml) warm water

1 pound (454 g) sausage, casing removed, or ground pork seasoned with a pinch of freshly ground black pepper

1 1-pound (454 g) package dried fusilli, penne, or rigatoni pasta

½ cup (120 ml) whole milk at room temperature

1 tablespoon (6 g) grated Parmesan cheese, plus additional for garnish

Heat the olive oil in a heavy-bottomed, 14-inch (35 cm) sauté pan over medium heat. When the oil is hot, add the leeks and cook, stirring occasionally, until wilted, about 5 minutes. Season with the salt. Cook 5 minutes more, stirring frequently to prevent sticking.

ଔ

Add the water to the leeks. Use your hands to break the sausage into small chunks and add them to the pan. Cook over low heat, stirring frequently, until the meat is cooked, about 7 minutes.

ଔ

While the meat is cooking, bring a large pot of water to a rolling boil, salt the water, and add the pasta; cook until al dente, using the package instructions as a guide.

ଔ

Drain the pasta and add it to the pan with the leeks and sausage. Add the milk and cheese and toss over medium heat for 1 minute. Serve immediately.

FEBRUARY

February is a fairly quiet month on the farm. It's still cold outside. I spend my time in the office, putting together information about the farm and confirming reservations for the upcoming month. This is the time of year when agencies contact us to arrange for groups from Italy, other European countries, and the United States to visit the farm and attend wine and food tastings.

The major holiday in February is Carnival. Every Sunday kids get dressed up in all sorts of costumes and go into the piazza and run through the streets of San Gimignano, throwing confetti at passersby. Even more festive are floats that make fun of politicians. Every Sunday for a month we eat fried pastries in the form of *cenci* and *frittelle*.

I'm in the kitchen, still licking sugar off my fingers after eating a rice fritter, when Mimosa arrives. Mimosa is one of our main employees and well known as "the life of the party." She and I are the same age and have a lot in common, including a passion for shopping. When Mimosa has gone on a shopping spree, she usually tells me all about it at work the next day, and often she brings things in to show me.

Mimosa starts telling me about going shopping at the market that morning. She turned to me and said, "Sarah, look at my nice my new bra!" Since there were no men around, she lifted up her shirt so I could see. At that exact moment, Fabio, a worker in his fifties, walked in. Mimosa screamed and pulled down her shirt. "Oh, *mamma mia*," said Fabio. Then he added, "Starting tomorrow I don't want to work in the vineyards anymore. I want to work here in the kitchen with you!

Fabio is a well-known character at Poggio Alloro, where he is famous for two things:

first, for being a hard worker who can do almost anything and, second, for being a man who loves women.

One day Renzo's attractive wife, Sandra, came to the farm on one of her rare visits. Sandra has a restaurant of her own that keeps her very busy, so we don't see her often. She's in excellent shape and wears her hair in a blonde bob, and she dresses very stylishly. Fabio couldn't help but sit up and take notice.

Later that day Fabio came into the kitchen looking for cleaning supplies, and he paused to chat with everyone. To Mimosa he said, "You can't imagine the knockout I saw today—a beautiful blonde with really beautiful curves! Who is she? Do you know her?"

Mimosa thought for a moment about who the mysterious blonde could be, and then she said, "You moron! You know who that was? The wife of Renzo, your boss!"

Fabio went pale and begged Mimosa not to mention this to Renzo. Mimosa couldn't keep it to herself, though, and pretty soon she had filled Renzo in on the story. Renzo and Mimosa decided to play a joke on Fabio. When Fabio went out to his car at the end of the day, he found an envelope on the windshield. Inside it was a note that read: "Effective immediately, you are fired due to sexual harassment in the workplace!"

Fortunately, Fabio knew immediately that it was a joke. He ran into the kitchen, shouting in his heavy Tuscan accent, *"Ragazzi, vu' siete proprio stronzi!"*—which means "Guys, you really are a bunch of jerks!"

The American

There are certain things that I always heard around our house. But I didn't give much thought to these pieces of information; I took them for granted. One of those things was that the farm across the road from ours was owned by *l'americano*, "the American." "The American" is a man named Howard Jelleme who had owned San Lorenzo for about twenty-seven years.

I knew this man's name, but I had never met him in person. Years ago he sold the property and returned to the United States, where the rest of his family lived.

The beginning of the friendship between my family and Howard started when he decided to move to Italy to buy the San Lorenzo estate but was still traveling back and forth to the States, so he made arrangements with my dad to take care of and cultivate his land until he came back to Italy. With any contract or formal agreement, Amico and Howard made all their promises with a handshake in the field. It was their style. And for many years it was like that.

But this year I received an e-mail from him saying he'd be in Italy for a week to take care of some paperwork and he'd like to have dinner at our farm with all his former partners and employees at San Lorenzo from over the years. I'm happy that he'll come to visit us, and I can't wait to meet him since I've heard about "the American" for so many years.

When the evening of the dinner arrives, he walks in and I say, "Oh, finally I get to meet the American!"

I find that he's very much like my father. They are the same height—or maybe I should say the same shortness—and have the same white hair. They even wear similar eyeglasses.

Amico comes into the office, and Howard sees him and with his charming American accent says, "Oh, Amico! How nice to see you!" The two shake hands and hug. They haven't seen each other in a long time. They remain locked in their embrace for a while in a clear sign of how pleased they are to see each other. Looking at them, I can't help noting more and more the startling resemblance—both are stocky and have the wrinkled skin of men who have done a lot of hard work in the fields.

After a wonderful dinner where we relax and chat and I get a chance to know Howard a little better and hear about his vineyard in New Hampshire, the evening comes to an end.

These two old friends are in the office saying good-bye. Howard says to Amico, "Hey, Amico, do you remember when a handshake was as good as a written contract?"

I'm standing across the room, watching them, and in that simple question I hear their whole world. It's the world of two men from another time. It would be hard to find

Clockwise, from upper
left: Tosca in the kitchen.
Mimosa setting up. Bruce
and Renzo. Vilma and
Giancarlo. Tommaso.
Michalea at work. Gena,
Mimosa, and Sarah. Sarah,
Mimosa, and Giancarlo.

Frittelle di Riso di Rosa

Rosa's Rice Fritters

Serves 6 to 8

6 cups (1.5 L) water

1 pinch salt

1 tablespoon (10 g) grated orange zest

1 teaspoon (3 g) grated lemon zest

2½ cups (475 g) Arborio rice

½ cup (50 g) plus 1 tablespoon (6 g) powdered
   sugar

½ teaspoon (2 ml) vanilla extract

2 eggs, separated

1½ cups (170 g) unbleached all-purpose flour

Olive oil for frying

The night before you plan to serve the fritters, place the water in a pot and bring to a boil. Add the salt and citrus zest, and then stir in the rice. Cook until the rice is very tender and has absorbed all the water, 50 minutes to 1 hour. (Add more water if the rice begins to look too dry before it is fully cooked.) Remove from heat and set aside to cool overnight.

The next morning, add the ½ cup (50 g) powdered sugar, vanilla extract, and egg yolks to the rice and stir gently with a spoon. Whip the egg whites to a soft peak and gently fold in the rice mixture. Gradually fold in the flour, about ½ cup (75 g) at a time.

In a pan or a pot with low sides, heat the olive oil to 350 degrees F (175 degrees C). Form the fritters into small balls by pressing two tablespoons together; gently drop them into the hot oil. Fry for 2 minutes per side. Remove with a slotted spoon and drain briefly on butcher's paper. Serve warm sprinkled with the remaining 1 tablespoon (6 g) powdered sugar.

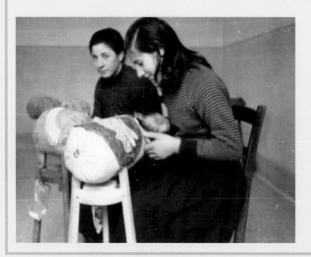

Rosa, left, learning ricamo a tombolo,
a traditional way of making lace.

the equivalent of these two in today's world—two men of honor who made sacrifices and worked hard, men whose word was golden.

Recently my dad and I went to visit Howard in New Hampshire, and he drove us to see his little farm with his little vineyard. What a surprise for us when we arrived! We saw a big sign on the wall of the house that read " San Lorenzo"—like his Italian estate. We were invited for dinner, and he offered us the last bottle of San Lorenzo, *italiana* estate, red wine, which he had been saving for special friends.

Making Prosciutto

Amico begins working on something that will probably keep him occupied for the rest of the afternoon: carving a prosciutto that weighs thirty-seven pounds (17 kg) and has been aging for a good twenty-four months.

Pigs are used to make a variety of products, including salt-cured Tuscan prosciutto. During butchering, the pig's rear haunches and its shoulders are removed and set aside to be made into prosciutto. First they are trimmed of excess fat. The outside of the prosciutto remains covered in rind, while the inside attached to the thigh bone consists of fresh meat. The round end of the femur pokes out from the meat. The prosciutto from the haunch is considered better than the prosciutto from the shoulder because it has a higher percentage of lean meat than fat, while the reverse is true for the shoulder.

At this point, a prosciutto weighs fifty-five to fifty-seven pounds (25 to 26 kg). First the prosciutto is covered with a paste of garlic and red wine vinegar so that the meat will better absorb the salt. Then it is placed on a wooden shelf in a salting room and covered completely with sea salt. A weight is placed on top of the meat to press it and draw out its liquids. This process helps the fresh meat absorb the salt and allows the resulting prosciutto to be stored for a long time.

After three days, the prosciutto is massaged to help expel any blood remaining in the aortic vein; it is then returned to the room. After thirty to thirty-five days, almost all of the salt covering the prosciutto has been absorbed, and any remaining salt is removed. The prosciutto is then rinsed with red wine vinegar and dried, and then the meat around the joint is coated with ground chile pepper. All that's left to do at this stage is to hang the prosciutto in a cool, dry place. Though modern drying environments with controlled temperature and humidity do exist, the best place to age a prosciutto is still the oldest room in the house, the one with a fireplace.

After a month in this room, the prosciutto is taken down and placed horizontally on a board for the final stage. The portion of the prosciutto that is not covered with rind is now coated with a plaster of untreated pork fat and black pepper. Once this plaster has dried, ground black pepper is spread over the surface.

This mixture is left to dry for a couple of days, and then the prosciutto is hung in the meat-curing room, where it will age for ten to sixteen months, depending on the size of the prosciutto. The more a prosciutto weighs, the longer it needs to be aged.

Amico goes down the three stairs to the meat-curing room with a look of concentration on his face. He pulls a little tool from the pocket of his apron—it's a small, pointed horse bone.

He inserts the end carefully into a prosciutto and holds it there for a few seconds. Then he pulls it out and places it under his nose. He closes his eyes and sniffs—this is how he can tell whether the prosciutto has gone bad or developed an "off" odor. He performs the same process for each prosciutto in the first stage of aging as well.

Then he selects a prosciutto that's one of the heaviest we've ever made, one that has been aged for a full two years. You can tell from the way he handles it that he can't wait to taste the first piece and then share it with the entire family and our guests.

Aptly describing the attention and care that he lavishes on this prosciutto is hard work—almost as difficult as making the prosciutto itself.

For his work he uses three or four different knives, each one with a distinct type of point and blade. He begins by carving the prosciutto into three parts. Then he removes the rind and the peppered plaster from each portion so that he can slice them. An electric slicer's rotating blade produces heat and changes slightly the scent of any cured meats it cuts, so the best way to cut prosciutto is still the traditional method—by hand. First the prosciutto is placed on a concave holder made from an olive branch. Then a long, sharp, flat-bladed knife is used to cut the best and thinnest slices possible. Of course, when you have a large number of slices to cut, an electric slicer is a necessity.

It's almost impossible to buy a prosciutto that tastes as good and has been aged as long as those made by Amico. Usually a prosciutto that you find for sale weighs from twenty-two to twenty-eight pounds (10 to 13 kg) and has been aged for only a few months, often using products that are far from natural. That's what today's market, accustomed to instant gratification, demands.

Who wants to wait twenty-four months to eat a slice of good prosciutto? Well, I do!

Arista di Maiale con Carote, Funghi e Uva Passa

Pork with Carrots, Mushrooms, and Raisins

Serves 6

2¼ pounds (1 kg) center-cut boneless pork loin

Sea salt to taste

½ cup (120 ml) plus ¼ cup (60 ml) extra-virgin olive oil

3 large carrots, chopped, about 1½ cups (200 g)

2 cups (480 ml) hot water

3 cups (225 g) sliced white mushrooms

1 cup (240 ml) vernaccia wine, or another dry white wine (such as sauvignon blanc)

1 cup (150 g) raisins

Season the pork with sea salt. Heat ½ cup (120 ml) of the olive oil in a heavy-bottomed, 4-quart (4 L) braising pan. When the oil is hot, add the meat and carrots. Cook over medium-high heat for about 10 minutes, turning the meat to brown it on all sides. Add the hot water, cover, and cook over medium heat for about 45 minutes.

℃

While the meat and carrots are cooking, heat the remaining ¼ cup (60 ml) olive oil in a separate 12-inch (32 cm) skillet over medium heat. When oil is hot, add the mushrooms and cook until the mushroom liquid evaporates, about 5 minutes; set aside.

℃

Transfer the mushrooms to the braising pan with the meat; and the wine and raisins. Cook, uncovered, for about 5 to 7 minutes to evaporate the wine slightly; then cover and cook for an additional 30 minutes, or until the pork is very tender. Add sea salt if needed.

℃

To serve, slice the meat as desired and transfer to a serving platter. Pour the sauce and vegetables over the top and serve at once.

Opposite: Luciano Giannini with Amico.

Tiramisù Poggio Alloro

Poggio Alloro Tiramisu

Serves 8.

3 eggs, separated

¼ cup (50 g) plus 1 tablespoon (14 g) sugar

8 ounces (240 g) mascarpone cheese

1 cup (240 ml) strong espresso coffee

1 cup (240 ml) hot water

38 to 40 ladyfingers, preferably Vicenzi brand

Cocoa powder as garnish

Using electric mixer, beat the egg yolks and ¼ cup (50 g) sugar together at high speed until mixture is fluffy and pale lemon yellow in color, about 5 minutes. Add the mascarpone and beat at high speed until very well blended, about 3 minutes; set aside.

ॐ

Using electric mixer, beat the egg whites in a clean bowl until they form medium-stiff peaks. Gently fold the whites into the egg yolk mixture; set aside.

ॐ

Combine the espresso with hot water and the remaining 1 tablespoon (14 g) sugar. Quickly dip the ladyfingers into the espresso. (Dip them no longer than a couple of seconds; otherwise, they will fall apart.) Arrange a single layer of the ladyfingers in the bottom of a deep-dish 8½-inch (21 cm) pie pan. (Or use individual custard dishes or martini glasses.) Cover with a portion of the mascarpone mixture. Continue layering the ladyfingers and the mascarpone mixture until the pan is full, ending with the mascarpone mixture.

ॐ

Refrigerate until well chilled. Before serving, scatter some of the cocoa powder over the top and cut into wedges.

MARCH

March 20 or 21 marks the first day of spring. When I was little, I didn't know that spring began on a specific date. After all, I'd been taught that spring has arrived when the swallows return to build their nests in the eaves and when daisies begin to grow in front of the entrance to the grotto.

The grotto was dug into a damp slope, so even in the middle of summer, when it was very hot outside, inside the grotto it was cool. We stored wine there as well as fruits and vegetables, such as watermelons and cantaloupe. The only problem was that we kids were all afraid of the grotto—it was so dark and deep that you couldn't see the bottom, and the key to the grotto was kept in a hole in the ground that seemed to be home to all kinds of spiders. Just to the left of the entrance, engraved in the tufa stone, were a pair of images of the face of the Madonna. When I was little, I thought

that a caveman had done them, but they were actually the work of my uncle, my mother's brother, a painter. They remained the same over the years without fading even slightly. He probably just quickly scratched them in the tufa with a sharp stick, but they were beautiful. After the especially rainy year of 2002, some earth slid down in front of the entrance to the grotto, so now there's a treasure buried there. Inside there are still two very old barrels of wine and some bottles of red wine (vintages 1970 and 1980) arranged in a neat tower.

The signs of spring are plentiful on the farm. The first tree to flower is always the almond tree that stands in front of the house on top of a small hillock. All it takes is the merest hint of warm weather and the tree's beautiful pale pink flowers begin to blossom.

When we were little, we didn't have a lot of toys. A ball was a real treasure, and if

we were lucky enough to get a ball, we were careful not to lose it. No matter how far it flew or rolled, we followed it and found it. Still, a new ball never survived more than a couple of days because I had a puppy, an Alsatian-German shepherd mix, that sooner or later would sink its teeth into any ball. Sometimes we resorted to crumpling up newspapers and then wrapping them in packing tape. This kind of ball didn't bounce, of course, but we could have fun with it just the same.

Our favorite game was making bows and arrows. My cousin Paolo designed them, and I constructed them. The quiver for holding the arrows was made out of newspaper and attached to the body using string. Arrows were made out of little reeds that we collected from a hut. The only problem was that the reeds actually made up the walls of the hut, walls that eventually were spotted with some large holes where we'd pulled out too many reeds. Our other specialty was to make go-carts that we then rolled down the hills.

We spent as much time constructing those carts as we did riding them, partly because after we'd hurtled down the hill at top speed a few times, the wheels and all the other parts of the cart would fly off. Typically, we'd lose the wheels, which, obviously, were a key element. We'd go up in the attic and look for old baby carriages or anything else that had wheels so that we could take them off and use them. Once we even used the wheels from my doll carriage.

Eventually, the time would come for *merenda*, or mid-afternoon snack. In a time before packaged cookies and chips, our *merenda* consisted of slices of bread drizzled with oil or tomato sauce, or zabaione. Often our parents left us on our own, as they were working in the fields, and on those days we made our own snacks.

Paolo had been taught that the zabaione was ready when it was the same color as a yellow Lego piece, so he'd spend forever beating the egg, holding a yellow Lego next to the bowl and waiting for the mixture to turn that same bright yellow color. Obviously, that never happened, and my sister Tiziana always had to go in and tell him to hurry up and come back out to play.

Marco and Paolo, my two cousins, are making all the wine boxes and checking the list of things they will need for the next day, such as bottles, labels, corks, DOCG labels, and so on. The next day they are going to bottle the vernaccia.

After that, Marco starts carefully checking the bottling machine to see if everything is working properly. The bottling machine was a great investment for the farm. Previously, we had to rent a truck that came and parked in front of the wine cellar to bottle the wine. This was pretty expensive and didn't give us the opportunity to bottle wine when we really needed to. Plus the fact we needed at least ten or twelve people to do the bottling.

To be bottled, the wine needs to be pumped to the bottling machine and through the filters to eliminate any sediments. Now we need just one person to handle the empty bottles and feed them into the machine; the machine does the rest, cleaning the bottle, filling it with the wine, corking it, and labeling it. The bottles come out ready to be put in a case. In total, now we need just four people.

Around this time of year, in February or March, we plant the tomato seeds that we've saved from the previous season. We always select seeds from the best tomatoes—those that are growing closest to the ground. (This low-growing position is where the tomato gets the most nutrition from the plant.) We start the seeds in little vases; when the seedlings have their first leaves, in April or May, they'll be transplanted in the garden, where the soil has already been enriched. When the tomato plants are six to eight inches (15 to 20 cm) tall, my father stakes them, tying them up with a willow branch. As the plants grow taller, he adds more willow branches for support.

Tending the Cows

At six o'clock every morning Umberto, the oldest of the siblings, goes out to the cowshed to do the milking. He takes his little wooden stool and his bucket and sits next to the cow. Milking is an art. It looks easy, but it's not. You have to grip the teat delicately and close it off at the top, then slowly slide your hand down to guide the milk to the exit.

Fresh milk comes out warm and strong-smelling. Early each morning Uncle Umberto brings a bucket full of milk to me in

Previous page: Umberto with the Chianina cattle.
Clockwise, from upper right: Bernardo. Giancarlo milking.
Amico.

the kitchen. You can drink it raw, but for the guests at Poggio Alloro we boil it to pasteurize it before serving it to them for breakfast.

The cowshed is a large building divided into two parts. On one side are the milk cows, which live in open spaces and can graze outside. On the other side are the Chianina cattle, both males and females. As the males are to be butchered for meat, they can't be allowed to graze freely, or they'll develop muscles and their meat will be too tough.

This side of the cowshed has a very simple cleaning system: There's a little canal with a conveyor belt that is turned on a couple of times a day. My cousin Marco and his grandfather, Umberto, use a pitchfork to toss the manure produced by the Chianina into this canal, and then the conveyor belt carries it all outside and deposits it in the manure heap.

Marco is tall, bright, and industrious, with a great sense of humor and a lot of curiosity and a willingness to work hard. After taking some time off from the university to decide what he wanted to do, he has worked full time for the business since mid-2008. Having studied winemaking, he is now our winemaker. His main job is overseeing the winery and the vineyards, but he's become our go-to guy for almost anything. Although we're cousins once removed, Marco and I were practically raised as brother and sister because we're only six years apart.

Marco had a very unusual "baptism" on the farm. All the children in my family spent their afternoons playing around the farm, especially in the cowshed. What could intrigue a child more than its ingenious conveyor belt? It's fascinating. Like a little river, it runs slowly and rhythmically, transporting excrement and hay out of the cowshed. Understandably, though, children can get a little distracted, and a distracted child might make a misstep and fall into the canal. That's just what happened to Marco: he found himself, not figuratively but literally, in deep sh——! I can't even describe how he shouted and cried afterward, nor can I count the number of baths it took to get that smell off of him.

We say it's good luck to step in manure, and if that's true, Marco must be the luckiest guy alive!

Umberto has lived his whole life with cows and calves. He doesn't need to rely on distinctive markings, no system of numbers and tags, to recognize his Chianina cattle. He knows each one's name and date of birth, the name of its mother, and even its personality. (I can only tell the males from the females; as my father says, "You wouldn't know a chickpea from a horse!") Yet our farm follows the

CHIANINA CATTLE

The Chianina breed is one of five Italian breeds raised for meat. The other four are Marchigiana, Maremmana, Podolica, and Romagnola. Chianina cattle are raised mainly in Tuscany, the Marche, and Umbria. The name derives from the Val di Chiana, an area of Tuscany south of Siena. This noble breed has ancient roots in the Etruscan age, and small bronzes from the Roman period depict Chianina cattle in bas-relief.

In addition to providing excellent meat, these cattle were traditionally used for plowing fields, pulling carts, and other field work. They were castrated at seven months, thereby becoming "oxen." These oxen grew to enormous size. At one year of age they were tamed, equipped with yokes, and introduced to light farming work. At about two years of age

they began to do heavier work. Yokes were balanced on their backs, and a rod was attached to each yoke. In turn, a coulter was attached to the rod. The oxen were guided by reins attached to nose rings. The nose ring was the ox's version of a horse's bit, as the ox's nose is his most sensitive spot. When they reached three years of age, the oxen were replaced by younger animals and slaughtered for their meat. The resulting meat was of high quality and very tender, due in part to the early castration of the animals.

In the 1960s oxen were replaced with tractors. Today castration is rarely practiced because it is considered cruel and extremely painful for the animals.

A bull named Donato from the postwar period holds the record as the heaviest bull ever raised in Italy. He weighed in at 3,924 pounds (1,780 kg).

standard practice of attaching a yellow plastic tag to the ear of each Chianina. This tag bears a code number that is linked to a kind of bovine passport containing all kinds of information about the animal.

My family has been working with Chianina cattle since we lived in the Marche region. In earlier days the animals' stalls were typically located on the ground floor of a house, with the bedrooms above. This meant that the heat produced by the animals rose to the upper floor and heated the rooms. Most houses had just one fireplace—two at the most—not nearly enough to heat the entire house. My father tells me that many relationships blossomed in the stalls, as unromantic a place as that might seem. In those days families gathered here after dinner because the stalls stayed so warm. The girls kept their hands busy spinning yarn, and the boys sat around and played cards. They would glance at each other across the room, and many a first love started that way.

Chianina beef is some of the best in Italy and, in my opinion, some of the best in the world. It's high in protein and low in fat, but it does have one downside: if overcooked, it becomes extremely tough.

In fact, the renowned *bistecca alla Fiorentina,* or Florentine steak, is always made with Chianina meat. Each steak is cut thick (four fingers high) and can weigh more than two pounds (1 kg). It must be grilled over a high flame, five minutes on each side, and seasoned with sea salt ground in a marble mortar and pestle. It is said that these steaks are cooked rare because that was the method used in the Middle Ages, but the real reason is that Chianina meat loses its flavor and turns much too chewy if it's cooked too long.

Every Saturday evening, as he has done for the last seventeen years, my father cooks *bistecca alla Fiorentina.* The steak is the centerpiece of a meal of hearty traditional Tuscan dishes, and it is always accompanied by an excellent red wine made of sangiovese grapes.

A bite of one of these phenomenal steaks washed down with a glass of Chianti is truly heavenly. The tannins in Chianti and in the region's other red wines give them the proper astringency to cleanse the palate and cut through the richness of the steak. This leads, of course, to taking another bite, and another sip of wine, and so on.

My father's recipe for beef tenderloin is another of his specialties and is always a great success with our guests. Sometimes I ask him to prepare beef tenderloin at four o'clock in the afternoon, and at eight o'clock he is still massaging it gently. He says it's important to rub it in a certain way in order to get the salt and pepper and the herbs deep into it. He really loves his meat, and I don't just mean eating it. Sometimes I think he's imagining

Clockwise, from right: Umberto and Giancarlo in the cow-shed. Bernardo rescuing a cow. Amico showing off Tuscan steaks before grilling. Amico helping guests with their selection.

that it's a woman, not a piece of beef. He rubs it and tenderly slices off any fat. Then he looks at it and turns it from side to side. Usually this drags on until I remind him, "Dad, there are guests for dinner. What am I supposed to serve them?" He's also very good at trussing the meat quickly.

Chianina cows also produce excellent milk, but they produce very little of it—barely enough to feed their own calves. It's rare, but sometimes the cows don't even produce enough to suckle their offspring. In these cases, Umberto bottle-feeds the calf, or one of our dairy cows serves as a wet nurse to the newborn. A Chianina cow would never nurse another cow's calf—she recognizes her own. But dairy cows, bred and raised for their milk, will do so. We always have one or two such dairy cows on our farm, and they are easy to recognize: they have brown or dappled coats and are short and squat, with very large teats.

When I was little, my own mother didn't produce enough breast milk to nurse me, and infant formula was hard to find in those days. Since we raised cows, a doctor suggested that she give me cow's milk, The cow whose milk nurtured me was named Nerina, "the little black one," because she was black, sweet, and very docile. Whole cow's milk would never be fed to an infant these days, as it's too rich and not very easy for babies to digest.

Below: Amico. Opposite, clockwise from upper right: Amico preparing the soil and planting seedlings. Bernado and Umberto. Umberto in the barn.

Spinaci Saltati
Sautéed Spinach

Serves 6 to 8

3 tablespoons (45 ml) extra-virgin olive oil

3 garlic cloves, peeled

2 (10-ounce) (560 g) packages fresh leaf spinach

½ teaspoon (3 g) sea salt, or to taste

Heat the olive oil and garlic in a heavy-bottomed, 12-inch (32 cm) skillet over medium heat. When the garlic begins to brown lightly and release its aroma, add the spinach and cook, tossing occasionally, for about 7 to 10 minutes, or until spinach is totally wilted. Add the sea salt and serve hot.

Spezzatino con Patate

Stewed Beef with Potatoes

Serves 6

3 tablespoons (45 ml) extra-virgin olive oil

2 pounds (900 g) beef chuck roast, trimmed and cubed

½ cup (125 ml) dry red wine, such as Chianti

3 garlic cloves, minced

1 onion, roughly chopped

3 carrots, roughly chopped

3 celery stalks, roughly chopped

½ cup (125 ml) beef broth

½ cup (125 ml) water

1 cup (225 g) tomato sauce

3 white potatoes, about 3 cups (450 g), peeled and cut into bite-size cubes

Sea salt and freshly ground black pepper to taste

Heat the olive oil in a heavy-bottomed, 5-quart (5 L) braising pan over medium-high heat. When the oil is hot, add the beef and cook, stirring often, until browned on all sides, about 5 to 7 minutes. Add the wine and cook, scraping bottom of pan, until the wine evaporates, about 10 minutes.

☙

Add the garlic, onion, carrots, celery, beef broth, and water; reduce heat and cover. Simmer slowly without stirring for 1 hour.

☙

Add the tomato sauce and potatoes, stirring to blend well. Season with salt and pepper. Cook covered for an additional 1 hour over low heat. Serve hot.

Torta della Nonna

Grandmother's Pie

Serves 8

2 cups (500 ml) milk

¼ teaspoon (1 ml) vanilla extract

2 eggs

½ cup (100 g) sugar

¼ cup (35 g) all-purpose flour

DOUGH

2 eggs

½ cup (100 g) granulated sugar

½ cup (50 g) plus 1 tablespoon (6 g) powdered sugar

2 cups (250 g) plus ½ cup (75 g) all-purpose flour

8 tablespoons (110 g) unsalted butter, melted and cooled

2 teaspoons (9 g) baking powder

¼ cup (40 g) whole or slivered skinned almonds

Powdered sugar

Preheat oven to 350 degrees F (180 degrees C). Spray a 9-inch (24 cm) springform pan with nonstick vegetable spray; set aside.

&

Combine the milk and vanilla in a heavy-bottomed, 2-quart (2 L) saucepan over medium heat. Heat until milk is warm, then reduce heat to low to keep the milk warm. Combine the eggs and sugar in bowl of electric mixer fitted with wire whisk beater. Beat at medium-high speed until mixture is fluffy and pale lemon yellow in color, about 5 minutes. Add the flour and beat just to blend. Slowly add the egg mixture to the warm milk, stirring constantly with a wooden spoon. Cook on low heat, stirring constantly, until mixture is the consistency of custard, about 5 minutes. Set aside to cool until lukewarm.

&

Make the dough. Combine the eggs, sugar, and powdered sugar in bowl of electric mixer fitted with wire whisk beater. Beat at high speed until mixture is creamy and pale lemon yellow in color, about 5 minutes. Add 2 cups (250 g) of the flour, a little at a time, and beat well after each addition. Add the melted butter slowly and beat at medium speed for 5 minutes. Add the baking powder; beat just to blend.

&

Scatter about ¼ cup (35 g) of the remaining flour on work surface and turn the dough out onto the surface. Top with the remaining ¼ cup (35 g) of flour. Using your hands, work the flour into the dough, kneading as you would bread dough, until the dough is no longer sticky. Divide the dough in half and roll one portion into an 11-inch (28 cm) round. Place the dough round in the prepared cake pan, patting gently to form the bottom crust. Press dough up the side of the pan, making a rim.

cx

Pour the custard into the pan, spreading evenly over the bottom crust and all the way to the edge. Roll the second piece of dough into a 9½-inch (25 cm) round. Place the second piece of dough over the custard to form a top crust and pinch the edges of the two crusts together to seal well. Scatter the almonds over the top and bake in preheated oven for about 50 minutes, or until crust is crisp and golden brown and the almonds are lightly toasted.

cx

Cool on wire rack for 10 minutes, then remove sides of pan. Cool completely and sprinkle with powdered sugar before serving.

Zabaione

Zabaione

This is a fairly substantial and sweet dessert, so the portions are small. When I made zabaione to determine the proportions (I'd never measured anything to make it before), it was as if we had all turned back into children again. Various members of my family came into the kitchen to watch, and we started talking about when we were kids and we'd make this and other simple things to eat as an afternoon snack. Then, when the zabaione was ready, nobody could resist! We gobbled up this delicious creamy treat.

Serves 4

4 egg yolks

4 tablespoons (80 g) sugar

2 teaspoons (10 ml) vin santo Toscano sweet wine

8 ladyfingers, preferably Vicenzi brand, for garnish, or substitute cat's tongue cookies or any type of plain cookie

Place the egg yolks and sugar in a medium bowl. Beat energetically with a spoon, always in the same direction, until the mixture is smooth and lighter in color, about 4 minutes. Add the vin santo and mix for 1 additional minute. (For a creamier texture, use an electric mixer.) Pour the zabaione into small cups and garnish with cookies. Serve immediately.

NOTE

Classic zabaione uses raw egg yolks, but you can gently cook the custard if you prefer. Place the zabaione in the top of a double boiler and cook, stirring constantly, over simmering water. Cook until thickened, but do not boil.

APRIL

In April the days get longer, and the weather grows more pleasant. Early in the morning the air is cool and refreshing, but after a few hours the temperature rises, and then it's a treat to be outside under the warming sun.

Easter (Pasqua), which arrives this year in April, is still a very widely celebrated holiday in Italy. Italians tend to travel over Easter vacation, so during this time the guest rooms at Poggio Alloro are full.

Pruning of the olive trees is finished around Easter, and a few olive branches are always set aside to be blessed in church on Palm Sunday. This ritual commemorates the arrival of Jesus in Jerusalem, where he was greeted by a crowd holding palm fronds. Since palm trees don't grow in our area, we use olive branches.

We also take hard-boiled eggs to the church to be blessed. The older women in my family always transported the eggs on a plate with a cloth napkin wrapped around it and knotted firmly at the top, and my grandmother did the same. After mass, they brought the eggs home and handed them out to all the family members to bring them luck. Then the eggs were eaten before lunch. The use of eggs at Easter has very ancient roots and stems from the fact that people were not allowed to eat meat or eggs during Lent, so when Easter arrived, they ate the eggs they had been denied. Eggs are also a symbol of birth and rebirth, and as such they represent the resurrection of Christ.

Over the last thirty years, real eggs have been replaced with chocolate eggs that come with all kinds of little toys inside. Such chocolate eggs are always beautifully wrapped, and

aprile

people even take them to church to have them blessed. They are not supposed to be opened until Easter Sunday, when with delight children crack them open and find out what's inside them. Since children are not the most patient creatures, a few chocolate eggs always end up being opened early. Whether on Easter or before, everyone enjoys indulging in the chocolate.

After Easter Sunday comes Pasquetta (Little Easter), which is what we call Easter Monday. Traditionally this is a day when families go on picnics if the weather is good. Easter Sunday and Easter Monday are both celebrated, of course, with food. In our family Easter Monday is not a day off. Instead it's the day when we start working the fields, planting corn and sunflowers. In fact, during this period we work harder than ever.

Tending the Chickens

Amico walks toward the henhouse, where he'll help my mother and my aunts care for the chickens. Rosa opens the henhouse, and thirty or so hens run out the door, happy to be free. They set about pecking the cornmeal and other grains that Amico places in their feeder. Then he rinses and refills their water containers.

While the hens are outside, Rosa collects the eggs and places them in a wicker basket to bring them back to the house. There are still two hens dozing on the straw in their nesting boxes, but as soon as Rosa gets close, they begin to squawk and cluck and beat their wings, as if they're shooing her away. As always, there are plenty of eggs—a couple dozen—and they're still warm. When Rosa exits the henhouse, she places the basket on

85

Maria Moschi and Rosa plucking a chicken.

the ground and Billo, our little black dog, immediately runs over to stick his curious nose into it and investigate.

Making Pasta

My mother brings the basket of eggs into the kitchen and takes out the flour, the rolling pin, and our wooden dough board. When Giannina is finished with the housekeeping, she comes down to give us a hand.

We each form a little volcano of flour on the board in front of us and break the warm eggs into the indentation in the center. We beat the eggs with forks until they're frothy and then begin to combine them with the flour. When the dough is crumbly, we start to knead. My mother's and my aunt's wedding rings glitter through the flour and bits of dough clinging to their hands. Their fingers have grown bigger and thicker over the years, and they can no longer remove their rings to do this kind of work. When they begin to roll out the pasta, the wooden rolling pins tick rhythmically against their rings.

When I was three or so, I used to watch with fascination whenever my mother made egg pasta. I'd stand behind her and observe that when she was using the rolling pin to roll out the dough, her thighs moved in concert with her arms. I had my own little board and tiny rolling pin, which had been my grandmother's, and after watching a while I'd climb up on a chair and stand next to her. My mother would give me a small piece of dough.

The process wasn't really clear to me, though. I'd put the dough on the board, then squish it with the rolling pin. Holding the dough still on the board, I'd move my thighs just the way my mother did, shaking my hips back and forth. My mother and my sisters always got a big laugh out of this.

I know how to make pasta now, but I still can't mimic my mother's movements exactly. My mother and my aunt wrap the sheet of pasta around the rolling pin and then quickly unfurl it, making the edge of the sheet of pasta smack against the board so quickly that it looks and sounds as if they're cracking a whip. I still try to imitate them, but I'm no good—the sheet of dough comes loose and the rolling pin slips from my hands and falls onto the floor. I think my mother still sees me as a little kid, shaking her hips and trying in vain to roll pasta dough. Oh, well—maybe the work is better left to the experts anyway.

Celebrating Easter

For the Easter lasagna, Mamma Rosa makes sure we start making sheets of pasta on Saturday night so they will have time to dry. At the same time, she makes meat sauce. The incredible smell of the sauce floats through all the

Pasta Fatta in Casa

Fresh Homemade Pasta

Serves 8

2 cups (220 g) semolina flour

2 cups (220 g) all-purpose flour

½ teaspoon (3 g) sea salt

6 eggs

1 tablespoon milk or extra-virgin olive oil, as needed

3 tablespoons salt (60 g)

Mix the flours and sea salt on a clean work surface; form into a tall mound and make a well in the center. Crack the eggs into the well and gently begin to beat with a fork. Mix a tiny bit of the flour with each stroke (this requires patience and cannot be rushed). As the dough begins to thicken and most of the flour has been blended, begin to stir the dough. If the dough seems to be too dry, add the milk or olive oil. Mix the remaining flour into the dough and work with your hands, kneading until dough is firm and elastic. Divide the dough into two balls, cover with a clean towel, and allow to rest for about 20 minutes.

☞

Lightly flour work surface with all-purpose flour and roll out one portion of dough into a large, paper-thin rectangle. Be sure to roll the dough thin enough. Lightly flour the surface of the dough, then roll it up, starting with the side nearest you. Repeat with the remaining ball of dough.

☞

For lasagna pasta, cut the dough into strips 4½ inches (12 cm) wide and as long as the pan. Unroll the strips and let dry for 2 hours. For tagliatelle, cut into strips about ⅜ inch (1 cm) wide; for tagliolini, cut into strips about ⅛ inch (3 mm) wide. Unroll the strips and toss the pasta with a small amount of all-purpose flour; let dry for about 1 hour.

☞

Bring a large pot of salted water to a rolling boil, add the pasta, and cook about 5 minutes. Drain well and use according to pasta recipe.

Sarah with a batch of fresh pasta.

rooms of the house and reaches the top floor and even outside. Anyone in the house during that time is constantly salivating.

The lamb, however, is Amico's responsibility. When my parents were young, their families kept sheep and goats, but we don't raise those animals anymore on our farm, so we buy a lamb from another organic farm. Amico spends hours butchering it; then he carefully seasons the leg of lamb with a mix of herbs and spices and cooks it in the wood-burning oven. The recipe for roasted lamb (*angello al forno*) is all his, but I do know that he uses some wild fennel to tone down the strong flavor of the lamb. He also sets aside the ribs, which my mother will fry up for another meal.

This morning I see him inspecting the oven. He lit the fire last night, so he's checking to see that it has reached the right temperature. He adds one more log to the fire and then collects all the pans that will go into the oven. There's a little buzz in the air as the dishes are prepared and cooking gets underway. Rosa has finished the lasagna, and the lamb and the potatoes are ready for the oven as well. Family members—some clearly coming straight from church in their fancy clothing and carrying eggs and olive branches—are in and out. A few pause to give Amico advice about how to cook the lamb. Still others ask if lunch is ready or come into the kitchen and search for food to pop into their mouths, though the women will chase them away. Children are everywhere underfoot, chocolate smeared around their mouths. In the midst of all this happy activity, Amico opens the door to the wood-burning oven and the scent of Easter is everywhere. Time to eat.

Lasagna is a very rich dish with a long history in Italy. The version with saffron is particularly special and is made by dissolving saffron in the béchamel sauce. The saffron lends the sauce a beautiful golden color as well as its characteristic scent and taste.

When I was little, our relatives from the Marche would come to visit over the holidays. After we had eaten a big meal, in the afternoon we'd play *bestia,* a card game that is usually played only by men while their wives clean up in the kitchen. Here's how the game is played. A player who has the winning cards or a good hand knocks on the table with his fist to indicate that he's calling the hand. My father let me sit next to him on the bench and knock for him. During play, everyone, large and small, was silent, and then after someone knocked and the hands were revealed, they all began to comment and tease and make jokes.

When the card games were over, we went outside and played bocce. My relatives from the Marche brought their bocce set, with its blue and red balls and one white ball. We played on the road by the house. First

Lasagne allo Zafferano
Saffron Lasagna

Besciamella is a light white sauce that comes from France (béchamel), but according to recent research, besciamella may have already been in use in Tuscany as salsa colla ("glue sauce") and imported to France by Catherine de' Medici. Besciamella has always been used in this lasagne dish to make it creamy and well blended.

Serves 6 to 8

3 (8-ounce) (225 g) packages oven-ready lasagna pasta, preferably Barilla brand

1 recipe Poggio Alloro Beef Ragu Sauce

Grated Parmesan cheese

BESCIAMELLA

5 cups (1.2 L) whole milk

½ teaspoon (0.25 g) saffron threads

5 tablespoons (75 g) unsalted butter

½ cup (75 g) all-purpose flour

1¼ teaspoons (7 g) sea salt

Make the Besciamella. Heat the milk and saffron in a heavy-bottomed, 3-quart (3 L) saucepan over medium heat just until moderately warm; set aside and keep warm. Melt the butter in a heavy-bottomed, 4-quart (4 L) saucepan over medium heat. When the foam subsides, add the flour and stir to blend well. Cook, stirring constantly, for about 5 minutes, or until flour is cooked but not browned. Slowly add the warm milk and saffron mixture, whisking to blend thoroughly. Bring to a full boil, whisking constantly, and cook until thickened and silky smooth, about 15 minutes. Add sea salt. Remove from heat and set aside; cool to room temperature. (Any leftover sauce can be frozen for future use.)

 C3

Preheat oven to 350 degrees F (175 degrees C). Spoon about ½ cup (120 ml) of the Besciamella into a 13 × 9-inch (33 × 23 cm) baking dish, spreading it to cover the bottom. Then spoon ½ cup (120 ml) of the Poggio Alloro Beef Ragu Sauce on top of it. Swirl the sauces together to blend slightly. Scatter about 2 tablespoons (25 g) of Parmesan cheese over the combined sauces.

C3

Arrange a single layer of the lasagna noodles on top of the sauce mixture, covering the entire dish. (You can cut the pasta with kitchen scissors to fit the pan.) Continue layering portions of the Besciamella, Beef Ragu Sauce, pasta, and cheese until the pan is full, ending with a layer of cheese.

❧

Bake the lasagna in preheated oven for 45 minutes, or until top is lightly browned and bubbly. Cut into squares and serve hot.

NOTE

If you use a deep-dish pan, you can make your lasagna up to 3 inches (8 cm) thick. Or you can make one 13 × 9-inch (33 × 22 cm) pan and one 8 × 8-inch (20 × 20 cm) pan. Bake both and give one to a friend, or freeze one for any easy dinner at a later time.

Fresh lasagna noodles can also be used in this recipe. Add to a large pot of boiling, salted water and cook for 2 to 3 minutes; drain well. Or you may also use lasagna pasta that needs to be precooked; just follow the package instructions.

FIORINI FAMILY'S EASTER MENU

❧

Antipasto
Mixed Antipasti
Mixed cured meats
Bruschetta

❧

First Course
Saffron lasagna

❧

Main Course
Roasted lamb
Beef tenderloin with herbs
Florentine beans
Roasted potatoes

❧

Dessert
Jam tart
Cantuccini *cookies;* vin santo
Coffee; grappa and limoncello

the white ball—the size of a golf ball—was thrown, and then the other balls, the size of apples but much heavier, were put into play. Sometimes I was lucky and had the honor of throwing the first ball. I was fascinated by the game. Actually, it was more of a solemn ritual than a mere game.

The goal of the game is to throw one of the balls as close as possible to the small white ball. You can also use your ball to knock away a competitor's ball so that it is farther away from the white ball. After each toss of a ball, there were shouts of encouragement and cautionary advice that grew louder and louder, interspersed with the sound of bocce balls hitting each other, which would cause the position of the balls to shift. Each player had his own style and method. Some threw right-handed with their left hands tucked behind their backs; others simply threw the ball overhand, and still others rolled the ball underhand and then gracefully slid one leg behind the other and bent over, like actors taking a bow before an appreciative audience.

A wonderful thing about my family's Marche roots is that it's given me a chance to taste traditional recipes from another region of Italy. The Monday after Easter, my mother cooks fried lamb ribs. As a good native of the Marche, she always serves the lamb ribs with cubes of fried custard and Ascolana olives, which are large olives that are pitted, stuffed with a meat mixture, and then fried. She fries everything in fresh lard, and it's all so good that as soon as she turns around, I grab whatever I can and stuff it into my mouth. Every time I do this, however, I find myself with a mouthful of burning hot food and I begin howling from the pain. Then she turns and sees me and scolds me. And I always say, "Mamma, you know that fried food tastes better when you steal it than when you eat it at the table. Everyone knows that!"

Clockwise, from left:
Marco and Tommaso.
Giancarlo hanging
the hams. Amico and
Billo getting firewood.
Vilma. Renzo. Paolo.

Ragù alla Poggio Alloro

Poggio Alloro Beef Ragu Sauce

This is a recipe for the basic meat sauce for pasta, lasagna, and other Italian dishes.
Any leftover sauce can be frozen for future use.

Makes about 8 cups

1 cup (240 ml) extra-virgin olive oil

3 carrots, cut into small dice

3 celery stalks, cut into small dice

1 medium red onion, cut into small dice

3 garlic cloves, minced

2 pounds (0.90 kg) lean ground beef

6 cups (1360 g) tomato sauce, preferably made from
San Marzano tomatoes

2½ teaspoons (14 g) sea salt

Heat the olive oil in a heavy-bottomed, 6-quart (6 L) saucepan over medium heat. Add the vegetables and garlic; cook until lightly browned, about 15 minutes. Add the ground beef and cook, stirring often to break up clumps, until meat is lightly browned, about 15 minutes. (The meat should be in small particles.) Stir in the tomato sauce and cook for an additional 25 minutes, stirring occasionally to prevent sticking. Remove sauce from heat and use as desired.

Piselli alla Fiorentina

Florentine Peas

*This is the perfect side dish to
serve with beef tenderloin.*

Serves 6 to 8

4 tablespoons (60 ml) extra-virgin olive oil

4 slices smoked bacon, diced, about 3 ounces (80 g)

6 cups (660 g) fresh shelled or frozen green peas

1 clove garlic, minced

½ teaspoon (1 g) minced sage

½ teaspoon (1 g) minced rosemary

¾ teaspoon (5 g) sea salt

1 pinch freshly ground black pepper

1½ cups (360 ml) warm water

Heat the olive oil in a heavy-bottomed, 14-inch
(35 cm) sauté pan over medium heat. When the
oil is hot, add the bacon and cook for 2 minutes,
stirring occasionally. Add the peas, garlic, sage, and
rosemary. Season with salt and pepper. Cook for
2 minutes, stirring frequently to prevent sticking.
Add 1 cup (240 ml) of the warm water and cook,
uncovered, until the water has evaporated, about
20 minutes. When all the water has evaporated,
add the remaining ½ cup (120 ml) warm water and
cook, uncovered, for an additional 10 minutes.

Vitello Arrosto

Beef Tenderloin with Herbs

A great alternative to the olive oil in this recipe is lard, which we use at the farm.

Serves 6

3 minced garlic cloves

1 sprig rosemary

1 sprig sage

2 pounds (1 kg) beef tenderloin

Ground black pepper to taste

Sea salt to taste

¾ cup (180 ml) extra-virgin olive oil or lard

2 cups (500 ml) good white wine, such as vernaccia or sauvignon blanc

Preheat oven to 400 degrees F (200 degrees C). Mince the garlic with the rosemary and sage leaves. Tie the beef with twine (see note) and sprinkle it with salt and pepper. Rub the beef with the herb mixture. Pour the olive oil in a roasting pan, add the beef, and roast in preheated oven. Every 5 to 10 minutes, pour a small amount of the wine over the meat to keep it from sticking to the pan. Cook until medium done, about 40 minutes (the roast should have a pink center).

ক৪

Before serving, slice and add a little of the pan juices to the slices of beef.

NOTE

Meat is tied—also known as trussing—to keep it juicy. As it cooks, a thin crust forms on the surface of the meat, sealing in the juices and retaining flavor. Meat is also tied to keep herbs and other flavorings in place. After meat has been tied, sage and rosemary—or sometimes pancetta and other items—are often slipped under the string. Meat that has been tied holds together as it cooks, so it looks more attractive when it is sliced. This technique can be used for any type of meat to be roasted in an oven or on a spit—beef, pork, veal, chicken, or rabbit—but it must be boned first.

To tie meat, use thin white butcher's twine. First wrap the twine lengthwise around the rolled up meat and tie it at the top with a pair of knots. The twine should be tight enough that it won't slip off, but not so tight that it squeezes the juices from the meat. Then wrap the twine around the meat repeatedly, moving it down the piece of meat at intervals of 1 to 1½ inches (3 to 4 cm). When you reach the end, knot the twine end around the original lengthwise piece of twine. The first length of twine is like a spine down the center of the meat, and the other lengths are like ribs going across it the short way. Be sure to remove the twine before serving the roast!

Patate al Rosmarino
Roasted Potatoes with Rosemary

3 large white potatoes, about 3⅓ pounds (1.5 kg),
  peeled and cubed

4 small sprigs rosemary

3 large unpeeled garlic cloves, smashed with side
  of knife

Sea salt and freshly ground black pepper to taste

4 tablespoons (55 g) lard, preferably fresh
  (nonhydrogenated), or substitute
  4 tablespoons (60 ml) extra-virgin olive oil

Preheat oven to 400 degrees F (200 degrees C). Put
the potatoes in a 14 × 10-inch (36 × 25 cm) baking
dish. Scatter the rosemary and garlic on top of the
potatoes and season with salt and pepper. Cut the
lard into small pieces and scatter over the potatoes.
(If using olive oil, drizzle it over the potatoes.)

CB

Bake in preheated oven for about 15 minutes, then
stir well. Cook an additional 45 minutes, or until
potatoes are lightly browned. Serve hot.

Dolce allo Yogurt
Yogurt Breakfast Cake

Don't be put off by the olive oil in this cake—it gives the cake a lighter flavor and texture than butter would. You'll barely get any equipment dirty making this, as you use the empty yogurt carton to measure most of the ingredients.

Makes one 10-inch (25 cm) cake, about 10 servings

1 (4-ounce) (120 ml) carton plain or fruit-flavored yogurt

4 eggs

3 cartons (1½ cups) (300 g) sugar

4 cartons (2 cups) (240 g) unbleached all-purpose flour

1 carton (½ cup) (120 ml) extra-virgin olive oil

1 tablespoon (8 g) baking powder

Preheat oven to 350 degrees F (175 degrees C). Butter and flour a 10-inch (25 cm) tube pan. Tap pan to remove excess flour; set aside. Empty the carton of yogurt into a small bowl and thoroughly wash and dry the carton.

಴

In bowl of electric mixer, combine the eggs and sugar and beat at medium-high speed until lighter in color and thickened, about 5 minutes. Gradually add the yogurt, beating until well blended and scraping down sides of bowl between each addition. Gradually add the flour, ½ carton at a time, beating to mix well and scraping down sides of bowl between each addition.

಴

Add the olive oil and beat to incorporate. Scatter the baking powder on top of the batter and beat an additional 2 to 3 minutes to blend well.

಴

Turn batter out into prepared pan and bake in preheated oven until a wooden pick inserted in center of cake comes out clean, about 45 to 50 minutes.

಴

Cool on wire rack for 10 minutes; then turn cake out onto rack, top side up, and cool completely before slicing.

MAY

In May nature explodes with green and other colors, too. The wheat fields are green and stretch as far as the eye can see. In a couple of months the wheat will be ready for harvest.

The first spring rains cause an avalanche of fava bean pods to sprout. My father brings fava bean pods home in enormous wicker baskets and turns them out onto the table. And the party begins! The best way to eat fava beans is to consume them raw with bread, prosciutto, sheep's cheese, and a glass of good red wine. We usually end up stuffing ourselves. During the harvest period we always serve them as a component of the antipasto.

About fifteen days later, artichokes begin to appear. The first artichokes are delicious and so tender. They are definitely best served as part of *pinzimonio*. For this dish,

extra-virgin olive oil, salt, and a dash of lemon juice (to keep our mouths from turning black) are combined in a small terra cotta bowl, and we just dip the artichoke leaves into the oil mixture. We pull one leaf at a time off the raw artichokes and moisten the soft, fleshy part with the *pinzimonio* and then eat it. I always race to get to the artichoke heart, my favorite part.

When I was little, I'd sit to my father's left and we'd eat artichokes together. He always complained that I didn't make good use of the leaves and left half of the flesh on them. "Look how I scrape them clean," he would say, showing me how to do it with his teeth. Once we got down to the tender part that I loved, he'd distract me and steal the heart. Then it was my turn to complain. My father and I don't have the most

maggio

communicative relationship, but we've always exchanged these small gestures of affection. For example, at Christmas my father always buys me a special little rare plant, a different one each year. He brings it into my office and tells me he's bought it for everyone to enjoy there, but we both know that it's a gift for me. I keep it in my office and tell everyone my father bought it for me.

Artichokes, like ice cream and chocolate, are extremely difficult to pair with wine. Raw artichokes are even more difficult than cooked because they're full of tannins that coat the palate. During the artichoke season, which lasts six weeks, we pick hundreds and hundreds of artichokes every day. Amico insists that we cook as many different artichoke dishes as possible. But after weeks of artichokes—as appetizer, first course, second

course, and side dish—he asks me, "Can't you serve a few more artichokes?" And I joke, "Yes, Babbo, tomorrow I'm pairing them with milk and coffee for breakfast!"

Artichoke season is also the season for picking peas and asparagus. While Maria, Giannina, and Rosa shell peas, they talk about a television program they have seen the night before—who won what, how much they won, and so on. Amico appears and asks how tonight's artichokes will be cooked. I say, "Enough with the artichokes, Babbo!" And then it turns out I'm in luck—the women in the kitchen have cleaned the smallest artichokes, boiled them in vinegar, and put them in jars to preserve them. No more artichokes tonight.

Meanwhile, I wash the asparagus under the tap, break off the hard ends of the stalks, and begin slicing them. The women pass me a bowl of freshly shelled peas, and I start making a favorite recipe for risotto featuring seasonal vegetables.

When dinner is over and the guests are all socializing with each other, Mimosa comes into the kitchen and asks me a question: *"Ma noi ce l'abbiamo il caffè d'orso?"* "Do we have bear coffee?"

Mimosa is originally from Albania, and though she has worked with us for many years and has lived in Italy for sixteen years, she still has a bit of an accent that makes us chuckle, and she sometimes mixes up words. This leads to some puzzling but very funny conversations.

I'm certain I haven't heard her right. *"Cosa, Mimosa? L'orso?"* "What's that, Mimosa? Bear?"

"Yes, bear," she insists.

I'm confused. "I'm sorry, but we don't have bears. We have Chianina cattle."

Mimosa responds, "What are you talking about cattle for? They want *l'orso, caffè d'orso.*"

Finally I get it. "Oh, *caffè d'orzo!*" It's barley coffee, a popular caffeine-free alternative, that a guest wants. "Sorry," I tell her, "but we don't have that either."

Insalatina di Carciofi
Artichoke Salad

*This salad makes a great
appetizer or antipasto.*

Serves 6

1 pound (500 g) whole artichokes

½ lemon

3 tablespoon extra-virgin olive oil

½ teaspoon (3 g) sea salt

1 pinch freshly ground black pepper

1 tablespoon (15 ml) lemon juice

*1½ ounces (40 g) thinly sliced Parmesan
cheese*

Clean the artichokes, removing the hard outer
leaves. Cut off the tips. Leave at least 1½ inches
(4 cm) of the stems, but peel off the stringy
outer layer. Place the cleaned artichokes in
a bowl of cold water. Squeeze the juice from
lemon into the water and drop the lemon into
the water with the artichokes. Set aside for at
least 30 minutes.

Drain and thinly slice the artichokes
lengthwise to make about 4½ cups (400 g).
Transfer the artichokes to a bowl and dress
with the olive oil, salt and pepper, and lemon
juice. Toss to combine. Then arrange the
artichokes on a serving platter and top with the
Parmesan cheese.

Carciofi Fritti
Fried Artichokes

Serves 6

1 pound (500 g) whole artichokes, preferably small

½ lemon

1⅔ cups (182 g) unbleached all-purpose flour

1½ cups (360 ml) cold sparkling water

½ teaspoon (3 g) sea salt

Oil for frying

Clean the artichokes, removing the hard outer leaves. Cut off the tips. Leave about 2 inches (5 cm) of the stems, but peel off the stringy outer layer. Place the cleaned artichokes in a bowl of cold water. Squeeze the juice from the lemon into the water and drop the lemon into the water with the artichokes. Set aside for at least 30 minutes.

છ

Cut each artichoke lengthwise into quarters. In a medium bowl, combine the flour and sparkling water and whisk until perfectly smooth without any lumps. Whisk in the salt. Heat the oil in a heavy-bottomed, 12-inch (32 cm) skillet. Toss the artichokes in the batter until coated all over. Fry the artichokes until golden, about 8 minutes. Drain immediately on a platter lined with butcher's paper or paper towels and serve as an appetizer or side dish.

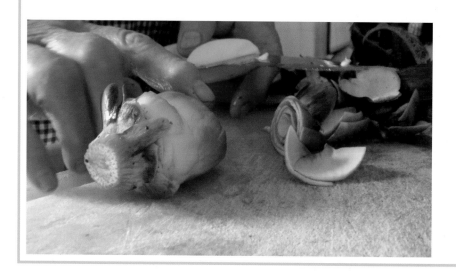

Tagliolini con Funghi e Carciofi
Tagliolini with Mushrooms and Artichokes

Serves 6

1½ pounds (680 g) whole artichokes

4 tablespoons (60 g) extra-virgin olive oil

1 tablespoon (5 g) minced parsley

2 cloves garlic, minced

3 cups (225 g) fresh mixed mushrooms (such as chanterelle, porcini, shiitake, and portobello), cleaned and sliced, or substitute frozen mushrooms, thawed and sliced

½ teaspoon (3 g) sea salt, plus additional for the cooking water

1 pinch freshly ground black pepper

⅔ cup (160 ml) warm water, plus additional as needed

1 generous pound (500 g) fresh tagliolini (see recipe for Fresh Homemade Pasta), or substitute packaged linguine

Clean the artichokes, removing the hard outer leaves. Cut off the tips. Leave about 1¼ inches (3 cm) of the stems, but peel off the stringy outer layer. Thinly slice the artichokes to make about 3½ cups (350 g).

♳

Heat the oil in a large nonstick pan and sauté the parsley and garlic over medium heat for 2 minutes. Add the sliced artichokes and sauté for an additional 3 minutes. Add the mushrooms, the ½ teaspoon (3 g) sea salt, and pepper. Stir continuously for a few minutes to combine, then add the warm water. Cook over medium heat, uncovered, for 20 minutes, stirring occasionally. If the pan starts to look dry and the contents start to stick, add additional water, 1 to 2 tablespoons (15 to 30 ml) at a time.

♳

Bring a large pot of salted water to a boil. Stir in the tagliolini and cook until al dente, about 4 minutes. When the pasta rises to the surface of the water, it's almost ready. Drain the pasta, add to the pan with the mushrooms and artichokes, and toss over medium heat for 2 minutes to combine. Serve immediately.

Risotto con Zafferano e Verdure Primaverili

Saffron Risotto with Spring Vegetables

Serves 8

4 tablespoons (60 ml) extra-virgin olive oil

½ cup (75 g) diced red onion

2 cups (300 g) diced zucchini

2 cups (300 g) sliced fresh asparagus spears, woody ends discarded

Sea salt and freshly ground black pepper to taste

4 cups (1 L) water

4 cups (1 L) vegetable broth

½ teaspoon (3 g) saffron

4 cups (760 g) Arborio rice

2 tablespoons (30 g) unsalted butter

Grated Parmesan cheese (optional)

Heat the olive oil in a heavy-bottomed, 14-inch (35 cm) sauté pan over medium heat. When oil is hot, add the onion and cook, stirring occasionally, until onion is wilted and transparent, about 5 minutes. Add the zucchini and asparagus, stirring to blend well. Season with salt and pepper. Cook for 25 minutes, stirring frequently to prevent sticking. Add ½ cup (120 ml) of the water as vegetables are cooking.

Combine 1 cup (240 ml) of the water with the vegetable broth and saffron in a heavy-bottomed, 3-quart (3 L) saucepan over medium heat. Cook to heat the broth mixture, about 10 minutes. Set aside.

Stir the rice into the vegetable mixture. Increase heat to medium-high and sauté, stirring, for 1 minute, then add ⅓ of the vegetable broth mixture. Reduce heat to medium, and cook, stirring, until rice absorbs the broth. Add 1 cup (240 ml) of the water, and cook, stirring, until it is absorbed. Continue to add broth and water, alternately, cooking and stirring, allowing the rice to absorb the liquid between each addition. When the rice becomes creamy, add the broth and water more sparingly, or the rice will become mushy. Cook until rice is creamy but still al dente, about 25 to 30 minutes total. Stir in the butter and scatter a little Parmesan cheese over the top, if desired. (Don't add too much of the cheese, or it will overpower the delicious taste of the saffron.)

Clockwise, from upper left: Giannina. Giancarlo in the garden. Amico and Billo in the olive grove. Tiziana in the kitchen with Rosa and Giannina.

Pasta all'Ortolana

Pasta with Garden Sauce

Serves 10

1 cup (240 ml) extra-virgin olive oil

1 cup (150 g) diced carrots

1 cup (150 g) diced celery

1 cup (150 g) diced red onion

1 cup (150 g) diced zucchini

1 cup (150 g) diced peeled eggplant

1 cup (150 g) green peas

1 cup (150 g) diced red bell pepper

1 cup (150 g) diced yellow bell pepper

2½ teaspoons (18 g) sea salt

2 (24-ounce) (680 g) bottles tomato sauce,
  preferably made from San Marzano tomatoes

1 cup (240 ml) water

35 ounces (1 kg) fusilli or penne pasta

Grated Parmesan cheese

Heat the olive oil in a heavy-bottomed, 4-quart (4 L) saucepan over medium heat. When oil is hot, add the vegetables and 1 teaspoon (7 g) of the salt. Cook, uncovered, for about 15 minutes, or until vegetables are wilted. If the vegetables begin to stick to the pan, add ½ cup (120 ml) warm water.

છ

Add the tomato sauce. Rinse each empty jar with ½ cup (120 ml) of the water and shake; pour into the pan and add the remaining 1½ teaspoons (11 g) salt. Stir to blend well and cook, stirring frequently, for about 20 minutes. Keep hot on low heat while preparing the pasta.

છ

Bring a large pot of salted water to a rolling boil and cook the pasta al dente, following the package instructions. Drain and toss with the vegetable sauce. Scatter Parmesan cheese over each serving.

Pollo alla Poggio Alloro

Roasted Chicken Poggio Alloro

Serves 8

1 frying hen, about 5 pounds (2 kg)

3 paper-thin slices pancetta

4 sprigs fresh rosemary

3 sprigs fresh sage

8 large garlic cloves, ground to a paste using a mortar and pestle

Sea salt and freshly ground black pepper

¾ cup (180 ml) dry white wine, such as vernaccia or sauvignon blanc

3 tablespoons (45 ml) olive oil

Grease a 13 × 9-inch (33 × 22 cm) baking pan with olive oil; set aside. Rinse the chicken inside and out under cold running water. Drain well and pat dry using absorbent paper towels.

Insert the pancetta, 2 rosemary sprigs, 1 sage sprig, and about 1 teaspoon (5 g) of the garlic paste into the cavity of the chicken. Using a small paring knife, cut six small slits in the skin of the chicken. Strip the leaves from the remaining 2 rosemary sprigs and mince with 6 leaves from the remaining 2 sage sprigs. Put a pinch of salt, a portion of the herbs, and a pinch of the garlic paste into each slit. Gently massage the seasonings into the chicken. Tie the legs of the chicken and place in prepared baking pan. Spread the remaining garlic paste all over the chicken and sprinkle with salt and pepper. Set aside to rest at room temperature for 30 minutes.

Preheat oven to 400 degrees F (200 degrees C). Pour the wine into the baking pan and spread the olive oil over the top of the chicken. Bake in preheated oven for 20 minutes, then turn the chicken onto its breast and bake an additional 40 minutes, or until well browned. Turn the chicken onto its back and bake an additional 30 minutes, or until a meat thermometer reads 170 degrees F (76 degrees C). Remove from oven and set aside to rest for at least 7 minutes before carving. Carve as desired and serve hot.

Rosa prepares chickens for the grill.

Guests enjoying the
garden and view.
Above: Giovanni.
Lower left: Sarah.

Panna Cotta con Salsa di Fragole

Panna Cotta with Strawberry Sauce

Serves 6

1 envelope (8 g) unflavored gelatin

2 tablespoons (30 ml) cold water

2 cups (480 ml) whipping cream

2 cups (480 ml) whole milk

½ teaspoon (3 ml) vanilla

½ cup (100 g) sugar

Mint sprigs for garnish

STRAWBERRY SAUCE

3 cups (450 g) fresh strawberries, hulled

Juice from half a lemon

½ cup (100) sugar

Stir the gelatin into the water in a small ramekin; set aside until the gelatin mixture feels spongy. In a heavy-bottomed, 3-quart (3 L) saucepan over medium heat, combine the whipping cream, milk, vanilla, and sugar. Bring the mixture to a boil, stirring to dissolve the sugar. Add the gelatin and cook until gelatin is completely melted. Be sure that there are no particles of unmelted gelatin. Divide the mixture among six 5-ounce (135 ml) ramekins and refrigerate at least 3 hours.

ෆ

Prepare the Strawberry Sauce. Combine the strawberries and lemon juice in work bowl of food processor fitted with steel blade and puree until smooth. Transfer puree to a heavy-bottomed, 3-quart (3 L) saucepan and whisk in the sugar. Simmer for 5 minutes, or until sugar is dissolved. Transfer to a bowl and refrigerate until well chilled.

ෆ

To serve, unmold the ramekins onto serving plates and top with Strawberry Sauce (or your favorite dessert sauce). Garnish with mint sprigs.

Opposite, lower left: Renzo.

JUNE

I wake to the scent of coffee wafting through the air and the sound of the stovetop coffee maker burbling on a burner. I hear Amico and Rosa talking quietly, trying not to wake me, but the light is shining through my bedroom window anyway, and I really want a cup of coffee with milk. The morning air is chilly, but in just a few hours it will be very hot. Yesterday my mother made *ciambellone,* and I'm going to eat a big slice of it with my coffee.

In the kitchen, my father pours me a big cup of hot coffee and then demonstrates for the millionth time that fresh milk that has been boiled makes the best cappuccino in the world. Billo the dog is sitting on my father's chair, behind him, with his muzzle poking out under my father's arm. He's trying to push my father's arm a little farther out to the side to get comfortable, and with his ears pushed back off his black face, he looks like a bat.

Amico quietly feeds him small pieces of cake dipped in milk and coffee. He didn't get the nickname Spoiled Billo for nothing—he won't eat his cake plain! If my mother sees my father giving cake to Billo, she'll take him to task for it. I see her turn and I begin to count down silently. Three, two, one. Here it comes.

"Amico, what are you doing?"

After a long day of work, I go outside for a walk. Nighttime is so relaxing and magical this time of year. The world is dark, but there are tiny lights teeming all around me. I've never seen so many—it's like a fairy tale. Fireflies arrive in the warm weather to light up the dark countryside, and crickets and cicadas play a lovely symphony. Fireflies are fascinating in the way they flash light to attract mates.

Often our guests will take a walk on the road in the evening to enjoy the cool air that arrives after dinner. One evening an

Australian woman came running up to me, all out of breath, and said, "There were two eyes following me in the darkness. It was very frightening! What kind of animal was that?"

I listed the possibilities: a wolf, or maybe a cat or dog.

"No," she insisted, "the eyes were at eye level. What a fright I had!"

For a moment I was puzzled, and then I thought of something. "Is it possible you saw two fireflies?" I asked.

"What are fireflies?"

In the years before we switched to organic farming, small insects such as fireflies, dragonflies, and butterflies were beginning to disappear, but now, fortunately, the countryside is filled with them again, and the ever-important balance between flora and fauna has been corrected.

In early June schools close for the long summer vacation. When we were little, vacation was a time when we took long walks through the fields and ran through the orchards looking for ripe plums or cherries. We hardly ever found any because we went over the same trees as often as twice a day— we didn't exactly give the fruit time to ripen! I followed my cousin Paolo everywhere and did everything he did and played every game he dreamed up, even the games that weren't considered appropriate for girls.

The days grew longer. Paolo, who was always full of sage advice, taught me that in summer it was okay to go barefoot. After that it was a struggle to get me to wear shoes of any kind. And it was even more of a struggle for my mother to wash my filthy feet at night—they were pitch black at the end of the day.

Once the corn in the field is ready, but

still young and tender, Amico picks a whole basket of ears and brings them to me in the house. If the fire is lit, we grill the corn, but otherwise we boil it and serve it with a pinch of salt. When the corn has grown taller, it's harvested using special machinery, and part of the still-green stem is fed to the cows. If you give a cow an ear of corn and then you try to pull it away from the other end, you won't succeed. Cows love corn.

The red wine, such as the Chianti, is now ready to be bottled, so Renzo and I, as always, must coordinate with Marco in the wine cellar to order the supplies we will need—bottles, labels, corks, and so on. This makes us dash like crazy from the farm to the wine cellar for a couple of days, but then we'll be ready for the bottling process. After we bottle the red wines, it's always good to wait at least three months before selling it to give the wine a chance to rest and mature in the bottle.

A Visit from Schoolchildren

Today is another warm June day. We've got one of our last visits from an elementary class before schools close for the summer. The children arrive promptly at 10:30 in their yellow school bus. Five- and six-year-olds step off the bus wearing brightly colored smocks and carrying satchels. I introduce myself, and I use the soft voice and simple words that help me engage more easily with kids of this age.

"First, we're going to see the bees and learn how they make honey," I tell them.

We walk in two lines along the road, and I can see they're already observing everything around them: Billo in the garden with Amico, all the vegetables growing there, the towers of San Gimignano in the distance, a white butterfly that's flitting around us. After we've walked for a few minutes, I point out to them how the natural environment is changing as we go from the countryside into the woods. In the woods we're in the shade. We follow the path that takes us to thirty or so hives. The hives are little wooden boxes where the bees live. Our beekeeper is there. For years we knew him as "the honey man," but his name is actually Piero. Wearing a yellow raincoat, Piero explains to the children that bees like yellow because it's the color of pollen, so it keeps them calm. He's not wearing gloves or a hat—the bees know him, and he's become immune to bee stings anyway.

I instruct the children to be very, very quiet and mime locking my mouth with a key and tossing it away. They all do the same. That way, in the silence we can hear the various sounds of the woods—birds chattering and the low buzzing of bees that are too focused on their work to notice us.

Piero lifts the top off of a hive and with

Clockwise, from upper left: Amico and Billo. Schoolchildren visiting the farm and tasting the fresh honey. Piero, the "honey man," hiding behind the bees and honey.

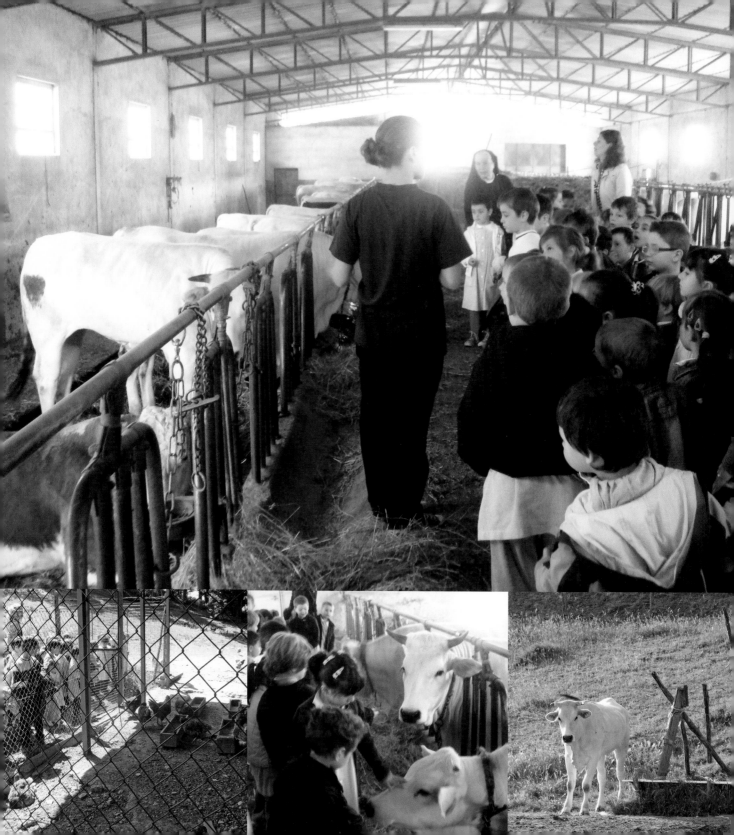

Ciambellone

Breakfast Cake

**Makes one 10-inch (25 cm) cake,
about 10 servings**

4 eggs

1½ cups (320 g) sugar

3 cups (360 g) unbleached all-purpose flour

2 sticks (220 g) butter, melted

¾ cup (180 ml) whole milk

2 teaspoons (10 g) baking powder

Preheat oven to 350 degrees F (175 degrees C). Butter and flour a 10-inch (25 cm) tube pan, tapping pan to remove excess flour; set aside.

☙

Combine the eggs and sugar and beat in an electric mixer at medium-high speed until pale lemon yellow in color and thickened, about 5 minutes. Reduce speed to low and add the flour, 1 cup (150 g) at a time, alternating with the melted butter. Beat to blend after each addition, scraping down sides of bowl. Add the milk and beat at medium speed for 3 minutes. Add the baking powder and beat just to blend well.

☙

Turn the batter out into the prepared pan, spreading it evenly. Bake in preheated oven for 50 minutes, or until a wooden pick inserted in center of cake comes out clean. Cool on a wire rack for 15 minutes, then turn the cake out onto the rack, top side up. Cool before slicing.

NOTE

This cake can also be made with cocoa. After the cake is mixed, remove 1 cup of the batter and blend it with 2 tablespoons (16 g) cocoa; mix well. Pour the white batter into the prepared pan and top with the chocolate batter. Continue with recipe instructions.

Fantasia di Crostini

Fantasy of Crostini

*Crostini are little slices of bread (French baguette) that are toasted on both sides.
You can then spread various toppings on the bread and serve them as appetizers or snacks.
The three colors of toppings reflect the colors of the Italian flag.*

Serves 6

2 baguettes, sliced about ½ inch (1.5 cm) thick

RED TOPPING

2 to 3 tomatoes, cut into small dice

Handful of fresh basil, torn into small pieces

Extra-virgin olive oil

Sea salt to taste

GREEN TOPPING

2 large garlic cloves, minced

1 cup (60 g) firmly packed flat-leaf parsley leaves

1 tablespoon (15 ml) extra-virgin olive oil

4 tablespoons (60 ml) mayonnaise

WHITE TOPPING

*5 ounces (125 g) Tuscan pecorino (pecorino
Toscano) cheese, or substitute 1 cup (100 g)
grated Parmesan cheese*

Extra-virgin olive oil

Preheat oven to 350 degrees F (175 degrees C).
Arrange the baguette slices on baking sheets and
toast in preheated oven until crisp, but not browned,
about 10 minutes. Remove from oven and set aside,
but leave the oven on.

To make the red topping, combine the tomatoes and
basil, drizzle with olive oil, and season with sea salt.

To make the green topping, combine the garlic,
parsley, and olive oil in work bowl of food processor
fitted with steel blade. Process until smooth and well
blended. Add the mayonnaise; process to blend well.

For the white topping, remove the rind from the
cheese and slice thin. Put the cheese slices on a
portion of the toasted baguette slices (or sprinkle
with the grated Parmesan); drizzle about 1 teaspoon
(5 ml) olive oil on top of each. Bake just until cheese
is melted, about 5 minutes.

Assemble the red crostini by spooning the desired
amount of red topping on some of the baguette
slices. Assemble the green crostini by spooning the
desired amount of green topping on some of the
baguette slices. Place all crostini on the same platter
and serve at once.

Maiale al Latte e Salvia

Milk Pork Tenderloin

Serves 8

6 large garlic cloves, ground to a paste using a mortar and pestle

2 pounds (900 g) boneless pork tenderloin, trimmed of silverskin

Sea salt and freshly ground black pepper to taste

¾ cup (180 ml) extra-virgin olive oil

1½ cups (360 ml) whole milk

6 whole sage leaves

Additional sea salt to taste if needed

Massage the garlic paste into the pork on all sides. Season on all sides with salt and pepper. Set aside for 30 minutes.

☙

Heat the olive oil in a 12-inch (32 cm) braising pan over medium heat. When oil is hot, add the pork and sear on all sides until lightly browned. Add ½ cup (120 ml) of the milk and the sage leaves; reduce heat to low and cook, covered, for 10 minutes.

☙

Turn the meat and add the remaining milk. Cook, uncovered, for an additional 20 minutes, or until an instant-read thermometer inserted in the thickest part of the meat registers 145 degrees F (62 degrees C).

☙

Allow the meat to rest for 10 minutes before slicing. Slice the pork ½ inch (1.5 cm) thick. Drizzle the thickened pan drippings over each portion and serve hot.

Insalata Verde

Green Salad

Serves 6

5 ounces (140 g) spring mix salad greens

3 Roma tomatoes, halved and sliced crosswise ½ inch (1.5 cm) thick

3 tablespoons (45 ml) extra-virgin olive oil

½ teaspoon (3 g) sea salt

Good-quality aged balsamic vinegar

Wash the greens and pat dry using absorbent paper towels. Combine the greens and tomatoes in a bowl and drizzle the olive oil over the top. Sprinkle with sea salt and drizzle a few drops of the balsamic vinegar on top. Using two large spoons, toss to coat all of the greens with the oil and vinegar.

Divide salad among individual chilled salad plates and serve.

Tortino di Verdure

Vegetable Tart

Serves 8

1 (8.65-ounce) (245 g) package frozen puff pastry,
  thawed

½ cup (120 ml) extra-virgin olive oil

1 red onion, diced

1 yellow bell pepper, diced

1 red bell pepper, diced

1 zucchini, diced

1½ teaspoons (9 g) salt

¼ teaspoon (0.2 g) freshly ground black pepper

4 eggs

¼ cup (25 g) grated Parmesan cheese

Preheat oven to 400 degrees F (200 degrees C). Spray an 8-inch (20 cm) springform pan with nonstick vegetable spray; set aside.

❧

Unroll the pastry sheet and shape it into a 13-inch (33 cm) round. Place the pastry in the prepared pan, gently patting it into the bottom of the pan and pressing the excess dough up the sides of the pan. Using a sharp fork, prick the pastry. Bake the pastry shell in preheated oven for 10 minutes, or until crisp and lightly browned. Remove from oven and cool. Reduce oven temperature to 325 degrees F (165 degrees C).

❧

Heat the olive oil in a heavy-bottomed, 12-inch (32 cm) skillet over medium heat. When the oil is hot, add the onion and cook until browned, about 15 minutes. Add the remaining vegetables and the salt and pepper. Cook, stirring frequently, until the vegetables are wilted, about 20 minutes. If the vegetables begin to stick, add ½ cup (120 ml) of water and cook until the water evaporates.

❧

Beat the eggs with the Parmesan cheese. Stir the vegetables into the egg mixture, blending well. Pour the mixture into the pastry crust. Bake in preheated oven for about 20 minutes, or until the filling is set. Remove from oven, allow to cool at least 5 minutes, and then remove the sides of the pan. Slide the tart onto a serving plate, slice into wedges, and serve hot.

JULY

As he does every day, my father got up at five thirty this morning and walked to the orchard, accompanied by Billo and carrying wicker baskets that he wove by hand this past winter. Attached to the handle of each basket is a piece of rope with an iron hook at the end. He climbs into a fruit tree, hangs a basket from a branch using the hook, and then tosses the fruit he picks into the basket.

At this time of year the trees are loaded with ripe, succulent fruit, which means they attract all kinds of birds. So for years my father has kept the birds away with his own scarecrow system: he places aluminum foil in the trees. The foil scares the birds when it reflects the light and moves in the wind. The trick works wonderfully, and the fruit is kept free from harm, even though the trees look as though they've been decorated for Christmas.

We farm organically. This means we don't use pesticides or other chemical products on our plants, which is a good thing. When bugs and birds do nibble on our fruit, we consider their bite marks a sign of quality. Surely they only go for the best fruit!

By the time my father fills his baskets with fruit, his brow is damp with sweat. Proud of his harvest, he heads up the road toward the house, faithful Billo still at his side. He wants me to make jam with all that fruit, so I end up in the kitchen surrounded by fruit, sugar, and many pots of jam on the stove.

Figs are always the most difficult fruit to pick. The fig tree has thinner, more brittle branches than other fruit trees, the source of the saying *il fico è traditore* ("the fig tree is a traitor"). Despite the fact that my father has

luglio

fallen from a fig tree three times (twice ending up in the emergency room), once again this year he decided to give it a try.

The fig tree—the same one—had no mercy on him. A branch suddenly broke beneath my father's weight, and once again he tumbled to the ground. We returned to the emergency room, where the nurses teased him gently that he ought to give up falling, as it wouldn't win him any prizes. Even an additional fall from a tree wouldn't earn him that trip to the Maldives, they giggled. When the fear had passed, we also laughed about it, though I did threaten to chop down that fig tree.

No matter. Two days later, there my father was again, balancing on those traitorous branches.

When the fig harvest ends, the plum harvest begins immediately, without a single day's pause between the two. We've made enough jam to last until next summer!

Bottling Tomatoes

My mother, Rosa, and my aunts Maria, Giannina, and Maria Moschi are busily preparing to bottle tomatoes. They tie on their aprons and cover their heads with handkerchiefs knotted at the back. The tomatoes, picked over the preceding two days, have been spread on large mats and left in the sun to ripen.

One of the four women washes the tomatoes, while two peel them, squeeze out their seeds, and then chop them. The last member of the team seasons the tomatoes with a little salt and basil and then begins to pack them into used wine bottles. A funnel is placed in the mouth of a bottle, and the tomato pieces

131

Giannina, Maria Moschi, Amico,
Maria, and Rosa bottling tomatoes.

are pushed through with a reed that has been whittled into two points at the end so that it resembles a pitchfork. Once a bottle has been filled, it is sealed with a cork. String is knotted around the cork to keep it from popping out as the bottle is being boiled. The final result looks something like a bottle of sparkling wine.

When all the bottles are ready, a fire is lit outside, and a large iron pot full of water is set over the fire. The bottles are placed in the pot horizontally, with layers of old newspapers packed in between them so that they don't break if they bounce around in the boiling water. Boiling the bottles both sterilizes them and creates a vacuum inside the bottles, which means their contents will last a long time.

I asked them to leave me a few of the less ripe tomatoes—I'm planning to use them tonight to make a nice *panzanella* bread salad for my guests. *Panzanella* is a humble Tuscan dish made of stale bread soaked in water, then squeezed dry and dressed. In fact, many traditional Tuscan recipes, especially those with rural origins, are made with stale bread.

You will find a lot of different combinations labeled as *panzanella* these days, but the classic *panzanella* bread salad consists of bread moistened and then squeezed dry, along with tomato, onion, and basil, seasoned with oil, salt, and a dash of red wine vinegar.

At Poggio Alloro our bread comes from a baker in Ulignano that we've nicknamed Marco Panaio, as if *panaio*, or bread baker, were his last name. Marco, sometimes known as Fischio, and his brother have supplied bread to farms and restaurants in and around San Gimignano for many years. Marco spends his days in a white van, delivering fresh bread. When we were little, he would bring us *schiacciata*, similar to focaccia, as an afternoon snack. He drinks a glass of wine at each farm that he supplies.

People tell the story that when Marco's nephew was fourteen, he tried to make the rounds with Marco, but he had to give up and go home mid-morning because he couldn't keep up with his uncle's drinking!

We always know when Marco Panaio is pulling up to the farm: first Billo barks, and then we smell the fresh bread when Marco opens the door to his van.

Once when my friend Johnnie Weber was visiting us from Texas and we were in the office, Marco finished making his usual daily delivery of bread and then said to me, "Are you ready to settle the bill now?"

I responded, "Of course!" I got out my calculator, and Marco told me how much bread he had delivered. As I hit the keys and paper scrolled out of the top of the calculator, growing quite long, Johnnie watched with a funny look on her face.

Finally, I pay up and Marco Panaio says good-bye and sets off for his next appointment.

After he leaves, Johnnie asks, "What's going on?"

I answer, "I was just paying the bread bill."

"How much was it?" she asks.

Offhandedly, I reply, "Six thousand euros." Johnnie was flabbergasted.

Then I add, "Well, it was for a year and a half!"

Johnnie couldn't believe it. The next day, when Marco made his deliver, she incredulously asked him how he could be so trusting and go so long without being paid for his bread.

Marco answered, "It's no problem. I know who's unreliable, and I make those people pay on delivery!"

Marco is also a truffle hunter. Once he shared a truffle with us, and we had a wonderful meal of tagliatelle with truffle that evening.

Threshing the Wheat

July brings the scent of wheat and toasted grains. My cousins Giancarlo and Paolo start their day early because this is threshing season. This key moment in country life occurs during the harvest and begins when the wheat (*grano*) is fully mature. Barley (*orzo*), oats (*avena*), and other grains were also threshed at one time.

It's already blazing hot outside. They'll work until noon, when they'll come back to the house for lunch and rest until at least four o'clock in the afternoon. It's normal to take a break like that when working in the fields. They begin working very early in the morning, before the sun comes up, and then they stop during the warmest time of the day and resume working in the afternoon, when the air is more breathable. They never know exactly what time they'll finish. They may work as late as ten thirty or eleven at night; after the sun goes down and a little breeze is stirring, it's easier to get things done.

At one time, threshing day was a day of celebration when the heads of the household wore their finest clothing. They awaited the foreman's arrival, and once he was there, threshing could begin. In those days it took twenty-five or thirty workers to handle the threshing; nowadays two people can do it on their own.

Pappa al Pomodoro

Tomato Soup with Basil

Serves 6 to 8

1 cup (240 ml) extra-virgin olive oil, plus additional for drizzling

2½ medium red onions, halved, then sliced lengthwise ¼ inch (6 mm) thick

10 large garlic cloves, peeled and minced

7 cups (1,587 g) tomato sauce

3 cups (750 ml) warm water

1 tablespoon (15 g) plus 2 teaspoons (10 g) sea salt

2 pinches freshly ground black pepper

½ teaspoon (1 g) crushed red pepper

½ pound (220 g) two-day-old Tuscan bread, sliced ½ inch (1.5 cm) thick and toasted

8 fresh basil leaves

Heat the olive oil in a heavy-bottomed, 6-quart (6 L) soup pot over medium-high heat. When oil is hot, add the onions and garlic. Sauté, stirring frequently, until onions are wilted and transparent, about 10 minutes.

❈

Add the tomato sauce. Pour the water into the sauce containers, shake to blend, and add the remaining tomato sauce to the soup; stir to blend. Add the salt and pepper. Cook for 15 minutes on medium-high heat, stirring to prevent sticking.

❈

Add the crushed red pepper and the toasted bread slices, pushing the bread slices down into the soup. Lower heat to medium and simmer the soup for 15 minutes. Using your fingers, tear the basil leaves into the soup; stir to incorporate, and cook an additional 5 minutes.

❈

To serve, ladle into shallow soup bowls, being sure to include some of the bread with each serving. Drizzle a small amount of extra-virgin olive oil on top of each serving and serve at once.

Opposite, above: Rosa, Maria Moschi, and Maria. Below: Gianina.

TUSCAN BREAD

Bread is a basic food in many cultures and has been since the medieval era. Back then, white bread and breads incorporating spices, herbs, or dried fruits were found only on the tables of the wealthiest members of society. Bread eaten by the common people was made with bran or "lesser" flours. Although flour may be made from any type of grain, by far the most common and most prized type of flour is wheat flour.

Tuscan bread is an excellent unsalted bread that lasts for several days, does not grow moldy, and remains intact. Because Tuscan bread is "flat" (unsalted), it goes well with Tuscan food, which is usually heavily seasoned. The lack of salt in Tuscan bread dates back to approximately A.D. 1100, when Pisa blocked the sale of salt to convince Florence, then its great adversary, to lay down its arms. The Florentines refused to accept Pisa's conditions and instead began to make bread without salt. It is also said that salt was very expensive at the time, so it was not used to make bread.

Farming families made bread only once a week, using a natural yeast starter (something like a sourdough starter) known as the "mother" (*lievito madre*), which was created by mixing water and flour together and allowing the mixture to ferment for a few days. Bread made with this kind of starter has a crisp crust with a tender inner crumb due to the microorganisms created through the fermentation process. The enzymes in the microorganisms cause the dough to rise during baking.

In those days, bread was stored in the home's *madia* along with the flour itself. A *madia* is a wooden

cupboard about forty inches (1 m) high with two doors at the bottom and an opening at the top. The top of the *madia* lifted up and was propped open with a sideways peg. Under the top was a wooden board where the bread could be kneaded.

Once loaves had been formed, they were placed on a table and covered with a wool cloth to keep them warm. They were left there to rise for about four hours before baking. While the bread was rising, the wood-burning stove was being prepared with kindling cut when the woods were being pruned. Different kinds of oak were added to the oven until the temperature reached 350 to 400 degrees F (180 to 200 degrees C). When the fire had burned out, the ashes were removed and the oven was cleaned; then a wooden peel was used to slide the risen bread into the oven, where it baked

Left: Tommaso.

for one hour at about 400 degrees F (200 degrees C).

At Poggio Alloro we named the farm's own line of grain and flour products La Madia in memory of my grandmother, who always kneaded bread inside this special cupboard.

The fresh bread stayed good for about a week inside the *madia*, but as the days passed it grew harder. Throwing away bread was out of the question (it was practically sacred), so it was soaked in liquid to soften it. Stale bread was cut into slices, moistened with wine, and then sprinkled with a few spoonfuls of sugar for a snack. It was also dunked into a cannellini bean purée for *ribollita* soup or served in a bowlful of *pappa al pomodoro*, a tomato soup. Really, the uses for stale bread were endless.

Pane Toscano

Tuscan Bread

Marco Panaio shared his bread recipe with me, but because he makes 1,000 loaves a day, it was difficult to scale down the recipe for just one loaf. So this recipe is from my translator, Natalie Danford, who is married to an Italian and makes this bread weekly. Any flour or combination of flours can be used as long as no more than one-third is a whole grain flour. Natalie uses one-third all-purpose flour, one-third bread flour, and, for the remaining third, an equal amount of rye flour and whole wheat flour.

Makes 1 loaf

¼ cup (60 ml) very stiff starter (see note)

1½ cups (360 ml) warm water

3 cups (360 g) flour

1 scant tablespoon (20 g) salt (optional)

In a large bowl, combine the starter and the water. Break up the starter with your hands or a spoon. Add the flour gradually and mix until a very shaggy dough forms. This is a fairly soft dough that firms as it rises, but if it seems too runny, incorporate about 1 tablespoon (8 g) flour at a time until it is firm. If it feels too dry, add about 1 tablespoon (15 ml) water at a time until it feels soft. Set aside to rest for about 20 minutes.

Knead the dough on an unfloured surface, using a dough scraper, until it is smooth and soft, about 10 minutes. Cut away a piece of dough (golf ball size) to use as a starter for your next loaf (see note). Tuscan bread typically has no salt; but if you prefer salt, add it at this point.

Place dough in a large bowl and set aside to rise. After 2 hours, gently pour the dough onto a floured work surface, sprinkle the top lightly with flour, and fold dough to incorporate air, then return dough to the bowl, smooth side up. Allow to rise until soft and puffy, about 2 additional hours.

Marco and Giannina.

Transfer dough to a lightly floured work surface and let it rest for 20 minutes. (If you want to make rolls or two smaller loaves, divide the dough now.) Shape dough into a tight round loaf, turning it several times on the surface. The top should be very smooth and the surface taut. Place the dough, smooth side down, in a basked lined with linen or a colander lined with a floured dishtowel; cover tightly with plastic wrap. Set aside to rise until almost doubled, about 3 hours.

Preheat an oven with a baking stone in it to 425 degrees F (220 degrees C) for about 45 minutes. Place a piece of parchment paper on a peel or on the back of a baking sheet. Flour the top of the bread, then hold the peel or baking sheet over the basket and invert the bread onto the parchment paper.

Working quickly (the dough will begin to spread and lose its shape), cut a deep cross into the top of the bread, then cut a diagonal line bisecting each quarter of the cross so that there are 8 slashes in the bread.

Bake in preheated oven until the bread is golden brown and sounds hollow when you tap on the bottom, 40 to 45 minutes. Cool completely before slicing.

NOTE

To make your first batch of starter, combine flour and water. Rye flour is particularly good for creating the first batch of starter. A fairly stiff starter seems to be more effective, meaning less of it is needed, so the bread doesn't develop the overly sour taste that sourdough breads sometimes do.

The night before baking bread, add about 1 tablespoon (8 g) flour and 1 teaspoon (5 ml) warm water to the starter and combine thoroughly. This mixture should become puffy and bubbly after a few hours.

If you know you won't be baking for a while, store the starter in the refrigerator in a clean glass jar with a tight-fitting lid. When you want to use it again, bring it to room temperature, discard any black material on top, and pour off any beery smelling liquid. Discard about half of the starter and refresh as above. You may need to repeat this for two or three days before you have an active, pleasant-smelling starter again.

This kind of starter needs to be refreshed every once in a while, but it will keep almost indefinitely.

First came the hard part: grain was threshed by hand with a scythe. It was bundled into sheaves *(covoni),* and then the sheaves were bound into large cones that would be left in the fields until mid-July. At that time, carts pulled by oxen were loaded up, and all the wheat was brought in and assembled into a single pile. The wheat was kept very dry, and a few days later the threshing machine arrived, pulled by a tractor.

Back then, the thresher was powered by a steam engine. First, firewood was gathered for the boiler. Water for the boiler was hauled to the farm in wooden barrels loaded on a trailer and pulled by oxen. When the threshing began, some people stood on top of the pile of wheat and tossed it down into the threshing machine, while others remained on the ground, where they collected straw with pitchforks and arranged it into piles. Threshing began as early as four or five o'clock in the morning, after a breakfast of coffee with milk and cookies. Then at eight o'clock a more substantial meal of beef stew and homemade bread was served, along with trays of cold cuts and plenty of water and wine.

After the men had been working for an additional hour and a half, girls came by with handmade baskets full of cookies and bottles of *vin santo.* The workers took turns breaking to eat. They often got fresh with the girls,

pinching them or furtively stealing cookies from their baskets, and the girls would get angry with them.

At lunchtime, threshing came to a halt and everyone gathered around a table in the shade. The meal was always a hearty one that included soup made with veal or chicken stock, and everyone ate a great deal. But the break lasted only a half hour or so, and then the work began again. At four o'clock the girls returned, bearing baskets of more homemade cookies and bottles of cold *verdea,* a sweet, unfermented wine with a greenish hue. It is made from a grape variety of the same name (also known as *verdicchio*). The grape must is filtered twice through a cotton sack to remove even the smallest amount of natural yeast that might cause it to ferment. Since *verdea* is non-alcoholic, even children can drink it, and it is appreciated by young and old.

At the end of the day a sumptuous dinner was served in the barnyard. Long tables were set up outside so that everyone could sit together and eat pasta with meat sauce and roasted chicken. After dinner there was plenty of dancing to music played on a barrel organ.

Once gasoline- and diesel-powered, self-feeding threshing machines and combines became the norm, these barnyard celebrations began to disappear, along with the laughter, the dancing, and the organ music.

Vegetables from the Garden

So life goes on as it does every day—though every day is special. I'm still here, thinking about what to put on this evening's menu. I have to go and take a look at the fresh vegetables that Amico has brought me from the garden. Something is sure to come to mind.

My father carefully arranges everything picked in the garden in wooden crates. It is displayed so beautifully that I'm almost sorry to move things around and mess up the image. Today in the crates I find beautiful yellow zucchini flowers and many summer squash. Another crate is full of eggplant, which I will definitely use to make eggplant parmigiana. To my surprise, off to the side I stumble across a crate of figs. Oh, no! More figs. I'll be making jam again today.

In the meantime, Amico, who knows it's almost lunchtime, has cleverly begun to cook eggplant and tomatoes in the embers left from the fire that was lit earlier to boil the bottled tomatoes. He's halved some eggplants and tomatoes and placed them in a small baking dish, then made little slashes in them that he fills with slivers of garlic and rosemary. He's laid thin slices of pancetta on top of the eggplant halves.

Amico never eats food that's been purchased somewhere, and he has no use for elaborate recipes that don't reflect his own background. I always admire that about him. Surely that's why in his mid-seventies he has the body and muscles of a man in his twenties.

I look around and think what a lovely job my father and my aunts have done with the flowers this year. There are beautiful terra-cotta pots with lemon trees and colorful flowers all over the terrace. In the center of all this plant life stand two large umbrellas that offer a spot of cool shade during this steaming hot summer.

I turn my gaze below me, to the field next to the stables. There are a number of wooden crates of potatoes there, and my cousin is hard at work, loading them into the van to take them to the farm. Bernardo has been busy all day long, harvesting potatoes in the field next to the saffron crop. The potatoes are stored in the crates for the rest of the year and used in many recipes in the kitchen and for the family.

Meanwhile, the Chianina cattle are grazing and enjoying this beautiful sunny day. They walk slowly toward the lake for a drink.

There are no loud sounds in the area, but the cicadas provide soft background noise. All is calm and quiet. As I turn toward my office, the sound of my steps causes the lizards who had been enjoying the sun to run frantically for cover, seeking out small holes in the wall where they can hide.

Marmellata di Fichi
Fig Jam

Makes 3 (16-ounce) (450 g) jars

2 pounds 3 ounces (1 kg) very ripe fresh green figs

4 cups (800 g) sugar

1 tablespoon (10 g) thinly sliced lemon zest

Trim the figs and cut off the harder stalk ends. Cut the figs into small pieces and place them in a deep pot with a heavy bottom. Stir in the sugar and lemon zest.

☙

Place over medium heat and bring to a boil. Cook over medium heat for at least 2 hours, stirring occasionally.

☙

Transfer the still-boiling fig jam to sterilized jars. Fill the jars, seal with tight-fitting lids, and turn jars upside down immediately to create a vacuum seal.

NOTE

This is how we preserve fruit at the farm, but you may want to process the jars in a hot water bath as an extra safety precaution. Select a large pot—it must accommodate the jars and enough boiling water to cover the jars. Screw on the lids securely and stand the jars on a rack in the pot, adding boiling water to cover the jars by at least 2 inches (5 cm). Boil 7 to 10 minutes. Carefully remove the pot from heat and remove jars (use a jar holder) to a flat area where they cannot be accidentally touched. Let them cool thoroughly; be sure the lid is concave, indicating a vacuum seal. The jam will store well for twelve to eighteen months.

Crostata di Marmellata

Jam Tart

Strawberry jam is traditional in this tart, but you can use any type of jam. You can also substitute vanilla or chocolate pastry cream for the jam.

Makes one 10-inch (25 cm) tart, about 8 servings

1 egg

⅓ cup (70 g) granulated sugar

6 tablespoons (45 g) powdered sugar

1 stick plus 1 tablespoon (125 g) butter, melted

1¾ cups (190 g) plus 1 tablespoon (8 g) unbleached all-purpose flour

1 tablespoon (15 g) baking powder

1 (12-ounce) (340 g) jar strawberry jam

Additional powdered sugar for sprinkling top of tart

Preheat oven to 350 degrees F (175 degrees C). Butter a 10-inch (25 cm) springform pan; set aside.

Combine the egg, sugar, and powdered sugar in bowl of electric mixer. Beat until pale lemon yellow in color and thickened, about 5 minutes. Gradually add the melted butter, alternating with 1¾ cups (190 g) of the flour and milk, beating to blend well and scraping down sides of bowl after each addition. Add the baking powder and beat to incorporate, about 2 to 3 minutes.

Lightly flour work surface and turn the dough out onto the surface. Scatter the remaining 1 tablespoon (8 g) flour over the dough. Knead until dough is firm (use additional flour if dough is sticky). Shape the dough into an even cylinder.

Cut off one-third of the dough and roll it into a circle about 10 inches (25 cm) in diameter and ¼ inch (6 mm) thick. Transfer the dough round to the prepared pan and press it firmly to form the bottom crust. Spread the jam evenly over the crust to within ½ inch (1.5 cm) of the edge.

Cut off half of the remaining dough and roll it into a rectangle about 6 by 9 inches (15 × 22 cm). Using a sharp knife or pastry cutter, cut dough into thin strips about ½ inch (1.5 cm) wide. Arrange the strips on top of the jam to form a lattice. Trim off strips at the edge of the pan.

&

Roll out the remaining dough into rectangle about 4 by 6 inches (10 × 15 cm). Cut into thin strips and cover the perimeter of the tart (the border without jam) and the ends of the lattice strips. Use your fingers to stick the strips of dough to the bottom of the pastry (if you moisten the dough with a few drops of water or milk, it may stick better).

&

Bake the tart in preheated oven until the crust is golden brown, about 35 to 45 minutes. Cool on a wire rack for about 15 minutes. Remove the sides of the pan. Just before serving, dust powdered sugar over the top of the tart and slice into wedges.

Sarah with fresh-baked jam tarts.

Mousse alla Pesca

Peach Mousse

Serves 6

1 envelope (8 g) unflavored gelatin

2 tablespoons (30 ml) water

2 cups (300 g) peeled, pitted, and chopped fresh peaches

½ cup (100 g) sugar

1 cup (240 ml) whipping cream, well chilled

Sliced peaches, blackberries, and mint sprigs as garnish

Stir the gelatin into the water in a small bowl. Set aside until the gelatin mixture feels spongy.

ଔ

Puree the peaches in work bowl of food processor fitted with the steel blade. Turn out into a bowl and stir in the sugar. Whisk to blend well and dissolve the sugar. In a 1-quart (1 L) saucepan over medium heat, combine the gelatin and ½ cup (120 ml) of the peach puree. Cook just to melt the gelatin, stirring often.

ଔ

Add the melted gelatin mixture to the remaining peach puree, combining thoroughly. Refrigerate until the puree is chilled but not set, about 1 hour. Whip the cream until stiff peaks form and gently fold it into the chilled peach mixture. Spoon portions of the mousse into small glasses or cups and refrigerate for about 4 to 6 hours, or until well chilled and set to a fluffy consistency.

ଔ

Garnish each serving with a peach slice, 2 or 3 berries, and a mint sprig.

August

The long, humid August days pass quickly. Every night a red-orange sunset streaks the sky. When I give tours of the farm I have to wear a hat to shield me from the sun, especially during the hottest hours of the day.

Today Amico brings an enormous basket of watermelons and cantaloupes to the house. It's so heavy that we have to leave it in the garden behind the house—even the two of us together aren't strong enough to lift it.

"Listen," he says to me as he picks up one of the watermelons and raps it with his knuckles, "this one is already ripe. I'll set it in the shade and we can eat it tonight."

That evening I try to pick up the watermelon to serve it after dinner. I try with both hands, but it's still too heavy. I roll it toward the house, and then I call Vilma to help me bring it into the kitchen. Once we've heaved it onto the table, I take out a knife so long that it looks like a sword and I start pretending to be a Japanese samurai, jumping around and making noises. I've got the knife up in the air, ready to bring it down and split the gigantic fruit in two, when Vilma says, "Quit clowning around! Let's try to serve the watermelon before dark."

"You're no fun," I complain, sinking the knife into the rind. It squeaks loudly as I cut halfway through it, revealing a bright red interior. My father was right—it's perfectly ripe. I

agosto

look around, and since I'm alone, I scoop out the center of the melon, the best part, leaving it hollow in the middle. I can't resist.

Vilma comes back into the room and, knowing me as well as she does, yells, "Sarah! Did you eat the center of the watermelon?"

With my mouth full of juicy, sweet fruit—so full I can hardly speak—I say, "Who? Me? No!"

Anyway, I know Vilma as well as she knows me—she wanted that piece for herself.

These days, my mom and dad, Maria Moschi, and Fabio, our helper, will be busy down in the saffron field digging up the saffron bulbs. They will be arranged in baskets and brought back to the farm to be sorted by size—small, medium, and large. Any that look sickly will be thrown away. Over the next few days, the soil will be prepared and sprinkled with manure. About ten days later, the field will be ready to be planted again with saffron bulbs.

Bernardo and Giancarlo are also busy harvesting the third cut (*il terzo taglio*) of the alfalfa, which is not easy work in this hot sun. Since alfalfa continues to grow after being cut, it can be harvested several times during a season. The first cut, which occurs in May, is the best one because it is richer in protein; the second cut is at the end of June, and the third one, which is in August, will be feed for the cows and rabbits. After this, the fields are plowed to get them ready for the November sowing.

Insalata di' Orzo

Barley Salad

The ingredients in this dish reflect the colors of the Italian flag.
Serve it as an appetizer or as the first course of a meal.

Serves 8

1 quart (1 L) water

1½ teaspoons (8 g) salt

9 ounces (270 g) pearled barley

3 Roma tomatoes, diced

4 ounces (120 g) fresh mozzarella cheese, diced

8 fresh basil leaves

2 tablespoons (30 ml) extra-virgin olive oil

Sea salt to taste

Bring the water and salt to a boil in a heavy-bottomed, 6-quart (6 L) saucepan over high heat. Add the barley and stir well. Cook according to package instructions, or until barley is tender. Drain in a fine strainer, run under cold water, and set aside to cool and drain.

☙

Combine the barley, tomatoes, and mozzarella in a bowl. Tear the basil leaves into the bowl with your fingers. Add olive oil and sea salt. Toss to blend well.

NOTE

Tearing the basil will retain its colors (it won't turn dark) and its pleasant fragrance.

Clockwise, from left: Renzo. Amico on his birthday. Amici del Chianti musicians Gino i' contadino and Cecco performing at the farm. Aunt Maria.

Bruschetta al Pomodoro

Bruschetta with Tomato and Basil

Serves 6

3 ripe Roma tomatoes, seeded and diced

Sea salt and freshly ground black pepper to taste

¼ cup (60 ml) extra-virgin olive oil, plus additional for drizzling

5 large basil leaves

1 loaf Tuscan or rustic-style bread, about 1 pound (450 g), cut into slices about ¾ inch (2 cm) thick

1 large garlic clove, peeled and sliced in half, plus additional as needed

Combine the tomatoes, salt, pepper, and olive oil. Tear the basil leaves into the mixture and toss to blend all ingredients. Set aside.

ଔ

Grill the bread slices over a fire or under a broiler until lightly browned and crisp on both sides. Rub one side of each bread slice with the cut side of the garlic. Season with salt and drizzle each slice with olive oil.

ଔ

Spoon a portion of the tomato mixture on each bread slice and serve.

Melanzane alla Parmigiana
Eggplant Parmesan

Serves 12 to 16

4 large round eggplants, about 3 pounds (1.4 kg), sliced lengthwise ½ inch (1.5 cm) thick

Sea salt

1 cup (150 g) unbleached all-purpose flour

3 eggs, beaten with ¼ teaspoon (1.3 g) sea salt

4 cups (1 L) canola oil

1 cup (100 g) grated Parmesan cheese

BESCIAMELLA

5 tablespoons (75 g) butter

½ cup (75 g) unbleached all-purpose flour

5 cups (1.2 L) milk

1¼ teaspoons (7 g) sea salt

¼ teaspoon (0.6 g) freshly grated nutmeg

TOMATO AND BASIL SAUCE

1 cup (240 ml) extra-virgin olive oil

2 cups (300 g) chopped red onions

6 cups (1360 g) tomato sauce

1½ cups (360 ml) water

1¾ teaspoons (12 g) sea salt

6 large basil leaves

Arrange a single layer of eggplant slices in a large colander set in the sink. Scatter sea salt over the slices to remove bitterness. Add a second layer and salt the slices. Repeat until all the eggplant has been salted. Place a dinner plate on top of the stacked eggplant, then place a heavy weight (such as a saucepan filled with water) on the plate to press the moisture out of the eggplant.

To make the Besciamella, melt the butter in a heavy-bottomed, 4-quart (4 L) saucepan over medium heat. When the foam subsides, add the flour and whisk constantly until mixture is well blended. Cook, whisking constantly, for 3 to 4 minutes. Add the milk gradually, whisking constantly. Bring to a boil to thicken. Whisk in the salt and nutmeg. Remove from heat and set aside; cool to room temperature.

To make the Tomato and Basil Sauce, heat the olive oil in a heavy-bottomed, 5-quart (5 L) soup pot over medium heat. When the oil is hot, add the onions and sauté until wilted and transparent, about 10 minutes. Stir in the tomato sauce. Add the water to the tomato sauce containers, shaking to blend, and add this mixture to the pan. Simmer, stirring occasionally, for 20 minutes. Tear the fresh basil leaves and drop into the sauce; stir to blend and cook an additional 5 minutes. Remove pan from heat and set aside.

Place a wire rack over a baking sheet and cover it with absorbent paper towels; set aside. Remove the eggplant slices from the colander and rinse briefly under running water. Pat dry using absorbent paper towels. Dip the eggplant slices into the flour, coating well and shaking off excess flour. Next dip them into the beaten eggs, coating well on both sides.

Heat the oil in a heavy-bottomed 14-inch (35 cm) skillet to 350 degrees F (175 degrees C). Fry the eggplant slices in batches as they are breaded. Cook on both sides, turning once, until golden brown and crisp, about 2 minutes per side. Using tongs, transfer the eggplant slices to the prepared wire rack.

Preheat oven to 350 degrees F (175 degrees C). Spoon ½ cup (120 ml) of the Besciamella into a 13 × 9-inch (33 × 23 cm) baking dish. Add ½ cup (120 ml) of the Tomato and Basil Sauce; stir gently to blend. Add a single layer of the fried eggplant slices. Cover with another layer of the two sauces, then scatter some of the Parmesan cheese on top. Repeat the layering, using all eggplant slices and ending with a topping of the cheese. Repeat with another 13 × 9-inch (33 × 23 cm) baking dish.

Bake in preheated oven for 40 minutes, or until bubbly and browned on top. Cut into squares and serve hot.

Nè di Venere nè di Martè,
nè si sposa nè si partè,
nè si da principio all'artè.
—Italian proverb

SEPTEMBER

September is a busy month. Our workers have to take care of spreading the Chianina manure in the fields to prepare for fall and spring planting.

Harvesting the Grapes

The old farmer's proverb quoted above means that you should never get married or start out on a trip or begin work on Tuesday or Friday. This saying stems from the fact that Tuesday, *martedì* in Italian, is named for the Roman god of war, Mars, and Friday, *venerdì,* is named for the goddess of beauty and love, Venus, who represents all women. Since war and women were traditionally seen as negative, those were considered unlucky days to start something. Most farmer's proverbs do have some scientific basis and aren't simply superstitions. They were a way of passing down information

orally, since farmers didn't have a written tradition or access to written sources.

I remember one year we young people of Poggio Alloro decided to start the harvest on a Tuesday. So one day in advance, my parents and aunts and uncles went into the vineyards with their shears and a few baskets and harvested alone for a little while—all to give the impression of starting the harvest on Monday for the sake of superstition.

The harvest usually begins between September 10 and 15, depending on the weather. Sometimes the grapes are even ripe a few days earlier. White grapes generally ripen before red, so they are always harvested first.

The first grapes that are ready for harvesting are the chardonnay, which are more delicate than other varieties and usually the first to mature; only a small percentage of the

farm's grapes are this type. Then it's time to harvest the vernaccia. Next, when they are very mature, we harvest the white malvasia and the San Colombano, which are used to make *vin santo*. Then we start harvesting the red grapes—a small percentage of merlot and cabernet, followed by the sangiovese, colorino, canaiolo, and ciliegiolo varieties. Even though it is a white grape, the trebbiano will be harvested last because it needs more time before it is ready.

The harvest begins early in the morning. We head out to the vineyard, armed with shears and baskets. We form two teams that line up on opposite sides of a row of vines, facing each other. Someone can always be counted on to say, "Be careful you don't cut off my fingers!" to the person facing him.

In the middle of the rows there's also a tractor with a cart attached behind it. The grape harvesters grab a bunch of grapes with one hand and with the other cut it off using the shears. Then they remove and discard any leaves and place the grapes in a basket. When the baskets are full, someone comes around and empties the grapes into the cart. When the cart is full, the tractor drives the cart to the wine cellar, where its contents are dumped into the hopper of a sorting machine. The tractor then returns to the vineyard. Meanwhile, the team of harvesters has moved down the row a few feet and started all over again.

When I was little, they used to lift me up into the cart so I could crush the grapes and make room for more. In fact, crushing grapes by stomping on them was a "technique" that was used for years to squeeze out the juice. Amico says that at one time young men and women would climb into enormous rectangular wooden vats and have a blast stomping grapes. The girls would knot their skirts above their knees—not too high, though, or people would talk—and occasionally the boys would down glasses of the grape must that trickled out of the tube as they worked. It was a lot of fun and a fitting way to celebrate the grape harvest. These days, though, this method is frowned upon, so presses are used instead. Crushing grapes with the feet can break the grape skins and set off the fermentation process too soon, when the grapes aren't yet in the vats, and that results in inferior wine.

In really warm years grapes may be harvested between five and eight o'clock in the morning. That way, the grapes are still cool when they arrive at the wine cellar, which means we don't lose the scent and we don't risk having the grapes start to ferment when they're still in the cart due to the high temperatures that occur once the sun is up.

As we're picking grapes, our two black dogs, Billo and his mother, Cila, run around us, and you can hear our laughter and shouts all the way back at the house. Harvesting grapes is noisy work—some of us have naturally loud voices, and some of us don't hear well, causing others to shout. It would be very unusual to witness a silent grape harvest. When tourists ask me where they can find my family members during harvest season, I always say, "Just follow the sound of their voices. You'll find them."

Making Wine

Bernardo is cleaning the floor of the wine cellar and working hard to make sure all the steel tanks are clean and ready for the new harvest. He spends most of his time in the vineyard pruning and taking care of every detail throughout the season, looking at the sky and hoping for another good harvest. A sudden change of weather can ruin the entire grape harvest, even in its last few weeks, and one full year of work will be wasted. Bernardo has two strong young guys to help him—Marco and his son Paolo—but they always follow Bernardo's advice because he has more experience with wine production.

Once the grapes are unloaded from the cart into the hopper, they are fed into a perforated cylinder that separates the grapes from the stems and stalks. This woody material is collected in bins and discarded. Meanwhile, the destemmed grapes are crushed—lightly if

Clockwise, from above: Bernardo. Bruce Weber. Maria Moschi. Paolo. Bernardo. Maria and Rosa. Opposite: Renzo selecting wine.

VERNACCIA DI SAN GIMIGNANO

The origins of the name vernaccia and the varietal itself are ancient. It is believed that the vernaccia grape was brought to San Gimignano from Liguria around 1200 by Vieri de Bardi. His descendants began cultivating the grape in San Gimignano and producing excellent wine from it. The name seems to come from the Latin word *vernaculus*, which means "from that place." This may explain why in Italy there are two other types of vernaccia in addition to the one from San Gimignano: Vernaccia di Serrapetrona in the Marche, a bubbly red, and Vernaccia di Oristano in Sardinia, a strong sweet wine.

The earliest known official document that mentions vernaccia is the 1276 tax code for the municipality of San Gimignano, which indicates that the market for this wine was already quite active in the thirteenth century. The tax code demanded payment of a duty of three coins for each shipment of vernaccia exported from the area.

In his *Divine Comedy*, Dante Alighieri wrote about vernaccia:

Questi, e mostro' col dito, e' Bonagiunta,
Bonagiunta da Lucca; e quella faccia
di la' da lui piu' che l'altre trapunta
ebbe la Santa Chiesa in le sue braccia:
dal Torso fu, e purga per digiuno
l'anguille di Bolsena e la Vernaccia.

This, and his finger then he raised,
Is Bonagiunta of Lucca; and that face
beyond him, pierced more than the rest,
holds the Holy Church in his arms:
he was of Tours, and purges by abstinence
Bolsena's eels and Vernaccia.
(Purgatory XXIV, 19–24)

Other texts also mention vernaccia. In a treatise entitled *Della natura dei vini e dei viaggi di Paolo III*, Sante Lancerio, the cellar master to Pope Paul III, recorded an order for eighty bottles of vernaccia, then noted regretfully that there were too few bottles produced to meet the request and that this wine was "the perfect drink for gentlemen, and it is a shame that this place does not make enough." A sonnet about the days of the week, written around 1300 by the poet Folgore da San Gimignano, refers to the wine in the verses on Wednesday: "Greek wine from the coast and Vernaccia."

Vernaccia was also on the table at the Medici-

Rucellai wedding in 1468 and on the tables of Lorenzo il Magnifico, who constantly requested that the town of San Gimignano supply him with the wine. It was at the home of the Medici that Pope Leo X grew accustomed to drinking vernaccia, and once he was installed at Saint Peter's he had it brought to him in Rome because he could no longer go without. Even Ludovico il Moro in 1487 ordered two hundred bottles to be served at his nephew's wedding. In 1695 Vincenzo Coppi, in his *Annali, memorie ed huomini illustri di San Gimignano*, described vernaccia as "a very delicate white wine, and one of the best and most admired wines made in Italy."

Today Vernaccia di San Gimignano is a DOCG wine with its own rules for production. It was also the first wine in Italy to receive DOC status in 1966.

The various abbreviations found on Italian wine labels refer to the governmental rules that cover wines:

DOCG (*Denominazione di Origine Controllata e Garantita*)—Certification of Controlled and Guaranteed Origin

DOC (*Denominazione di Origine Controllata*)—Certification of Controlled Origin

IGT (*Indicazione Geografica Tipica*)—Protected Geographical Indication

Of course, none of these terms can tell you whether you'll like a wine or not, but they do indicate something very important—the quality of the wine. These regulations establish precise rules for specific production zones, including the types of grapes used, vineyard practices, aging temperatures, and, most crucially, the yield per hectare, which is very important when it comes to quality.

For example, the grapes for Vernaccia di San Gimignano must be grown in the municipality of San Gimignano and nowhere else in the world. The grapes grow in soil that is a mixture of sand and clay and has a high fossil content. Maximum production is limited to ninety quintals per hectare.

Vernaccia is generally a beautiful pale yellow. Some versions, especially the Riserva, are almost golden. Riserva describes a wine that has been aged in wooden barrels; in the case of vernaccia, it is aged for at least twelve months.

Vernaccia isn't as fruity and floral as many other wines, such as chardonnay, but if you close your eyes and sniff a glass of vernaccia, you can't help but notice its unique and prominent bouquet. It is often described as having notes of green apple and ripe white fruits and yellow wildflowers, such as broom. It's definitely dry, with a slightly bitter

finish reminiscent of bitter almonds. The taste tends to linger on the palate. It also has a strong savory quality that it gets from the minerals in the San Gimignano soil.

Often when I talk about flowers or fruit in relation to wine, guests look a little lost. I try to explain to them that these sensations are very subjective, and the palate is very subjective as well. It's a good idea to close your eyes and think about what scents or flavors from daily life the wine brings to mind—like the smell of the jar of jam

you opened that morning, or the smell of flowers you got for your birthday, and so on. All you need to do then is to link these scents and tastes with what you're tasting at

that moment. Obviously, a wine expert is not made in a single day. You need to taste and taste and taste!

Vernaccia wine is best served chilled at 46 to 50 degrees F (8 to 10 degrees C) or, in the case of the Riserva version, at 54 degrees F (12 degrees C). Vernaccia goes well with certain dishes, such as fried baby vegetables or

cheeses that have not been aged for too long. A Riserva di Vernaccia matches well with more strongly flavored cheeses and spices, *ribollita*, spaghetti with seafood, risotto with artichokes, "white" lasagna with vegetables (and no tomato sauce), all kinds of grilled fish, rabbit braised in vernaccia, turkey with saffron, roast chicken with potatoes, and other white meats. And don't forget that vernaccia also makes an excellent aperitif!

A consortium dedicated to vernaccia, the Consorzio della Vernaccia di San Gimignano, was founded in 1972 to promote this special wine. Today the consortium goes by the name Consorzio della Denominazione di San Gimignano because it now covers other wines produced in the area (and not just vernaccia). It was created with the goal of protecting and supporting winemakers and playing several other key roles.

they are white grapes and more firmly if they are red grapes. A pump then moves them to a stainless steel vat.

At this point, the process for white grapes diverges from that for red grapes.

Making white wine calls for keeping the grapes for one night at 46 degrees F (8 degrees C) in the stainless steel vats. The next day the grapes are pressed with a "soft" press. The resulting liquid is placed in another vat, while the skins are delivered to a local distillery, where they will be used to make Vinacce di Vernaccia grappa. Italian law dictates strict control of grappa production, a monopoly of the state, so grappa cannot be made on the farm. But in our store we are allowed to sell grappa from our own grapes if it has been made at a state-authorized distillery.

At this point in the process, the liquid is known as must. The must is clarified at 50 to 54 degrees F (10 to 12 degrees C), as enzymes separate from the larger particles, causing them to sink to the bottom of the vat. One day later, the clear must is poured off into another vat, while the dregs (feccia) remaining at the bottom are also delivered to a distillery and made into grappa. The clear must begins to ferment at a controlled temperature ranging from 57 to 64 degrees F (14 to 18 degrees C). In some years, fermentation may take place at 54 degrees F (12 degrees C). This lower temperature requires a longer fermentation period but results in a more floral and fruity wine.

On average, white wine ferments for twenty days. Put simply, fermentation is the transformation of the sugars in alcohol and carbon dioxide by means of yeast. When all the sugars have had a chance to develop, the wine can be drawn off. As during the clarification process, the dregs are left behind in the vat. At this point, the must has become wine. It is drawn off into another vat, where it will age until it is bottled—usually in March of the year after the harvest. Any remaining dregs are again sent to a distillery to produce grappa.

Until just a few decades ago, white winemaking was so complicated that most farmers combined white and red grapes. As a result, almost no true white wines were produced, and the wine was of lesser quality overall. That didn't matter much, though, because vino del contadino ("farmer's wine") was made for home use.

As new winemaking technologies developed and knowledge about winemaking improved, this system was abandoned completely. In Italy today, white wine is made from green grapes only and red wine from red grapes only.

For red winemaking, the grapes are placed in a stainless steel vat at 82 to 86 degrees F (28 to 30 degrees C). Fermentation begins the next day. Unlike green grapes, red grapes are fermented with their skins. In fact, red grape skins are the source of the polyphenols that give red wine its signature aroma, structure, and color.

The presence of the skins during fermentation creates the so-called "cap," a layer of skins that rises to the top of the vat. This layer keeps the skins separate from the dregs of the must, which settles at the bottom of the vat. To encourage proper fermentation and greater extraction of color during the entire maceration process, which lasts about one week, repassing is performed frequently. Repassing encourages contact between the must and the skins and is performed using a pump that draws the must from the bottom of the vat and brings it up to the top, where the skins are.

Once the fermentation process is complete, the wine is drawn off. First, the must—which is actually wine at this point—is poured into another vat. Then the skins are pressed by a machine that extracts the last of the substances that help wine with its final phase. This extract is added to the previously

Insalatina Autunnale
Autumn Spinach Salad

Serves 6

5 ounces (140 g) baby spinach, washed and patted dry

5 ounces (135 g) (12 mm) Tuscan pecorino (pecorino Toscano) cheese, diced, or substitute Asiago cheese

4 tablespoons (80 g) pomegranate seeds

½ cup (75 g) raw pine nuts

6 tablespoons (90 ml) extra-virgin olive oil

Sea salt to taste

Divide the baby spinach leaves among 6 chilled salad plates. Divide the diced cheese, pomegranate seeds, and pine nuts into 6 equal portions and arrange a portion on top of each salad serving. Drizzle each salad with 1 tablespoon (15 ml) of the olive oil and sprinkle with sea salt.

Left: Rosa and Maria. Right: Amico, Billo, and Sarah. Below: Rosa, Maria, Bernardo, Giancarlo, Maria Moschi, Amico, and Sarah.

Patate alla Poggio Alloro
Potatoes Poggio Alloro

Serves 6 to 8

3 russet potatoes, about 3½ pounds (1.5 kg), peeled and finely minced

2 cups (480 ml) whole milk

1 cup (70 g) shredded Emmentaler cheese

1 cup (135 g) chopped red onion

2½ tablespoons (35 g) butter, melted

2 teaspoons (14 g) sea salt

¼ teaspoon (0.2 g) freshly ground black pepper

2 eggs, well beaten

1 cup (100 g) grated Parmesan cheese

Preheat oven to 375 degrees F (190 degrees C). Butter and flour a 13 × 9-inch (33 × 22 cm) baking dish; tap to remove all excess flour. Set aside.

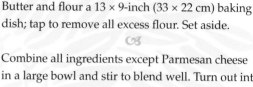

Combine all ingredients except Parmesan cheese in a large bowl and stir to blend well. Turn out into prepared baking dish and scatter the Parmesan cheese over the top. Bake in preheated oven about 45 minutes, or until set, browned, and bubbly on top. Cut into small squares and serve hot.

NOTE

The potatoes can be minced in a food processor, but be sure to drain off the liquid.

Umberto with potatoes.

poured-off wine. At this time, the second round of fermentation—known as malolactic fermentation—begins, a process that is fundamental for red wine. This is the phase in which the malic acid becomes lactic acid and the wine turns mellow, as more acidic notes are eliminated. Young red wines are left to age in stainless steel vats and bottled around June or July of the year following the harvest, whereas the reserve and selected red wines are aged in barrels—either a small type known as a *barrique,* which holds 59 gallons (220 L), or a large cask that holds about 528 gallons (2,000 L). Our choicest red wines are aged for an average of twelve months, while our vernaccia is aged for only three months. Some years, when the quality of the grapes is judged to be particularly exceptional, the vernaccia is both aged and fermented in barrels. This is the wine known as Vernaccia "Le Mandorle."

The barrels used at the farm are all made of French oak. In general, wood gives wine a deeper, more developed bouquet with subtle aromas of spices, licorice, vanilla, and so forth. Red wine aged in oak also develops better and the tannins are mellowed. Aged wine is usually more valuable due to the greater labor required and the cost of the wood. The barrels can be used for four years at the most. After that, they no longer lend their characteristic aroma to the wine. All our barrels are stored in a stone cellar that is twenty-six feet (8 m) below ground. The room remains at a constant temperature, winter or summer, and has an ideal level of humidity.

One of our most special wines is made in this room—our San Gimignano Rosso DOC Convivio. This red wine, made with 90 percent sangiovese and 10 percent cabernet sauvignon, is aged for twelve months in both casks and *barriques.* A greater quantity of wine can be aged in the casks, and the wine spends a longer period of time there. The wine spends less time in the smaller *barriques,* but it picks up a stronger aroma of wood from them and in general develops a stronger bouquet during that time. After aging for twelve months, the wine reflects both types of containers.

The word *convivio* derives from the Latin *convivium,* which means a banquet or symposium. In Greek and Roman times, a symposium was a convivial occasion where people gathered around a table and drank while singing songs, dancing, and conversing. That image always brings me back to the first day of my sommelier course. The lecturer that day told us that the best way to drink red wine is in a convivial manner, with family and friends. What better way to honor our red wine than to celebrate life by drinking it in the company of good friends and family!

Donzelle Fritte

Fried Dough

These savory balls of fried dough are sometimes served as an appetizer and sometimes in place of bread with a meal. For a really delicious treat, insert a small cube of pecorino or Parmesan cheese in the center of the ball before frying.

Serves 8 to 10 (makes about 20 pieces)

2 tablespoons (25 g) active dry yeast

1 cup (240 ml) warm water (105–115 degrees F) 40–46 degrees C)

3 cups (360 g) unbleached all-purpose flour

1 teaspoon (7 g) salt

2 tablespoons (45 g) lard

1 quart (1 L) oil for deep frying, heated to 350 degrees F (175 degrees C)

Stir the yeast and warm water together; set aside until yeast is foamy, about 5 minutes. Place a wire rack over a baking sheet and cover the rack with absorbent paper towels; set aside.

☙

Place the flour in a large mixing bowl. Add the yeast mixture, salt, and lard; mix well. Using your hands, knead the dough for 5 minutes. If the dough feels extremely sticky after kneading and clings to the side of the bowl, add additional flour, 1 tablespoon (8 g) at a time, until the dough is soft and malleable but not sticky.

☙

Lightly oil a separate large bowl. Form the dough into a ball and place it in the bowl, turning to coat lightly with the oil. Cover with a clean dishtowel and set aside in a warm, draft-free spot to rise until the dough is doubled in bulk, about 2 hours. (At the farm, we set the bowl near the radiator or on a sunny table.)

☙

Punch the dough down and divide it into 20 pieces; form round balls about the size of Ping-Pong balls. Fry in batches in the preheated oil, taking care not to crowd the balls. Fry the balls until browned, about 2 to 3 minutes. Working carefully so that you don't burn yourself, turn the balls with a fork and fry an additional 2 to 3 minutes. Using a slotted spoon or strainer, remove the balls and place them on the wire rack. Repeat with remaining dough balls. Serve warm.

Fagottini di Pecorino

Pecorino Cheese in a Pastry Box

Makes about 18 pieces

1 pound (450 g) hard pecorino cheese, rind removed

1 (17.3-ounce) (534 g) package frozen puff pastry, thawed

1 egg yolk, beaten

Preheat oven to 400 to 425 degrees F (200 to 220 degrees C). Line a baking sheet with parchment paper; set aside.

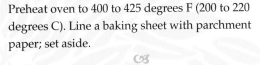

Unroll the pastry, and using a sharp knife or pastry cutter, cut the sheet into 3 × 3½-inch (7.5 × 9 cm) rectangles. (Use firm, precise cuts; don't crimp the edges or the pastry won't rise.) Cut the cheese into 1 × 1½-inch (2.5 × 4 cm) rectangles, about ¼ inch (6 mm) thick.

Place a piece of the cheese on one side of each pastry rectangle. Fold the uncovered portion of the pastry over the cheese and seal the edges like a piece of wrapped candy, gathering and pinching the edges to seal.

Place the pastries on prepared baking sheet and brush the tops with the beaten egg yolk, taking care that none of the egg drips onto the parchment. Bake in preheated oven until golden brown and puffed, about 20 minutes. Serve warm.

Torta di Mele

Apple Cake

The small amount of butter in this cake makes it a very light dessert. It may also be served warm for breakfast and goes well with a good cappuccino. This is made in the fall and winter, when local apples are abundant. I like it best with rennet apples, but you can try any apple you like, as long as they are nice and ripe with soft flesh, such as the Golden Delicious variety.

Makes one 8-inch (20 cm) cake, about 8 servings

3 eggs

1 cup (200 g) sugar

1½ cups (180 g) unbleached all-purpose flour

½ cup (120 ml) whole milk

2½ tablespoons (40 ml) butter, melted

2 teaspoons (10 g) baking powder

½ teaspoon (3 ml) vanilla

3 medium apples, about 1¾ pounds (800 g), peeled, cored, and cut into wedges ⅛ inch (3 mm) thick

Powered sugar as garnish

Vanilla ice cream (optional)

Preheat oven to 350 degrees F (175 degrees C). Butter and flour an 8-inch (20 cm) springform pan, tapping the pan to remove excess flour; set aside.

Combine the eggs and sugar in bowl of electric mixer. Beat at medium-high speed for 5 minutes, or until pale lemon yellow in color and thickened. Gradually add the flour, milk, and butter, stopping to scrape down sides of bowl after each addition. Beat at low speed until each ingredient is blended, then increase speed to medium high and beat for 3 minutes. Add the baking powder and vanilla and beat an additional 2 minutes to blend well.

Turn batter out into the prepared pan. Arrange the apple slices vertically, with the core side down, in concentric circles in the batter, beginning with the outside edge of the pan and continuing to the center. The arrangement of the apples should resemble a rose in full bloom.

Bake in preheated oven for about 50 minutes, or until a wooden pick inserted in center of cake comes out clean and apples are lightly browned.

Remove from oven and cool for 10 minutes. Remove the sides of the springform pan. To serve, cut the warm cake into slices, then scatter powdered sugar over each serving. Add a scoop of vanilla ice cream to each serving, if desired.

OCTOBER

On a day of heavy rain, my mother is bringing in some wood to build a fire. I go to help her, and my father, as usual, repeats the old proverb *Se vuoi vedere una donna buona a poco, mettila ad accendere il fuoco*, which means "If you want to see a woman be good-for-nothing, ask her to light a fire." My nephew, Ulisse, comes in and, smiling sweetly, asks, "Auntie Sarah, can I have hot chocolate for my afternoon snack?" Then Ulisse notices his Grandpa Amico cupping an ear to indicate that he didn't hear what was said, so Ulisse shouts loudly, "I said, 'Hot chocolate!'"

This is their favorite game. His grandfather asks, "What did you say?" and Ulisse shouts again, "Chocolate!"

His grandfather says, "Make yourself a nice sandwich with your grandpa's prosciutto instead of eating that junk."

While they have this exchange, they begin to chase each other around the table. Soon they're shouting, "I'm going to get you" and "No, you're too old, Nonno. You can't catch me."

"I'm coming to get you!" my father roars.

Our favorite game as children was played at the table, where my father let me drink a sip of his wine even when I was very small. He allowed me to lift the bottle and pour red wine into a glass. He showed me exactly how much I could drink, and he would hit the table with his hand to signal me to stop pouring. Once, to rile him up, I kept going and made him hit the table three or four times as I laughed and laughed.

Given that we're not too busy this afternoon, I start to make crepes and béchamel so that we'll have everything ready to make

ottobre

crespelle tomorrow. My mother taught me this recipe, and as soon as I start making the crepes I sense her standing behind me. I turn, and my mother pretends she wasn't observing me. I start working again, and she says, "Watch out, they're going to burn."

"No, they're not!" I say. "I'm keeping an eye on them. I won't let them burn."

She says, "Well, they burn in a second."

"Don't worry, Mamma. I'll take care of it."

I turn to open the refrigerator to take out a little butter to grease the pan, and when I turn back to the stove, she's there, flipping my crepes. "I told you, they burn in a second," she says.

I look at the crepes and they certainly don't look burned—just the opposite. "Mamma…" I say, and she responds, "A mother is like a record—she repeats the same

thing over and over." No argument there—she's said that very thing to me hundreds of times!

This is the season when most of the winter vegetables grow, and the kitchen is full of baskets of cauliflower, spinach, chicory, fennel, and radicchio, but all night I've been smelling something unusual in the kitchen. I stick my head out of my office and I see Aunt Maria, Aunt Giannina, my mother, Rosa, and my father, Amico, cleaning *cavolo nero,* or black cabbage (also known as Tuscan kale or lacinato kale); they're pulling off the tough ends of the stems and blanching the leaves in boiling water. Uncle Umberto's face appears at the glass door to the kitchen, and he reaches in to grab the bucket for milking the cow. My cousin Marco is getting in everyone's way as he makes himself a prosciutto sandwich.

Black cabbage is a typical Tuscan

vegetable with somewhat tough, wrinkled leaves and dark blue-green tips. This plant blooms only once, and when it does we have to collect the seed and replant it so that black cabbage will be available in all seasons. Since black cabbage is the main ingredient in the classic Tuscan soup *ribollita*, the dish can be made almost year-round.

Tuscan *ribollita* is possibly the best known of the many Tuscan soups, and everyone makes it in his or her own way. There are hundreds of different recipes in the various parts of Tuscany. There are, however, certain ingredients that must appear in any version: cannellini beans, black cabbage, Tuscan bread, and extra-virgin olive oil. The term *ribollita* means "reboiled" and refers to the fact that traditionally the soup was first boiled for several hours when it was prepared, and then it was reheated all week long. In fact, the more it was reheated, the better it got, because the bread absorbed the flavors and melded with the other ingredients. This is a farmhouse dish made by farmers who cooked with whatever was growing in their gardens and who survived largely on plant protein such as beans, as meat was rarely available to them. Bread was soaked in the soup when it was too stale and hard to eat

any other way. Even today, *ribollita* is made with bread that is a couple of days old.

Picking Mushrooms

With this rainy weather there are a lot of wild mushrooms sprouting, so Amico goes off to hunt mushrooms. My father is an excellent mushroom hunter and has his secret location. As a little girl, I would get to go with my father when he went mushroom hunting, since he was sure I could never find his favorite mushroom spot on my own. In fact, if I didn't hurry to keep up with him as we walked through the woods, I would quickly get lost. When Amico was young, he hunted truffles, too, but eventually he was forced to abandon that time-consuming pastime.

Sometimes he returns with wicker baskets full of mushrooms of all kinds—porcini, russula, ditole—as well as ovolo mushrooms, which are really rare and resemble eggs (hence the name). When they are small, they look exactly like peeled, hard-boiled eggs with an end cut off so that the round tip of the yolk pokes out! In a good season we eat mushrooms every day for a week, cooked in every possible way.

When I get to my parents' apartment

Clockwise, from right: Rosa.
Amico. Ulisse.

this evening, I find my mother cleaning a nice batch of mushrooms, all spread out on the table, with black soil still clinging to their stems and a few oak leaves stuck to their damp caps.

My father has lit the fire. Tonight for dinner we're going to eat three grilled porcini mushrooms, one apiece, and a nice piece of Chianina beef, also grilled. I spy a pile of ashes close to the flame, and I know that some onions are cooking underneath. My father digs into the ashes with a spade, then reaches in with his bare hands and picks up the onions and places them on a plate so I can remove the aluminum foil and the onion skins. I cut them and drizzle on extra-virgin olive oil, then sprinkle them with salt and pepper and a drop of red wine vinegar. We grill some bread, too, and I pour a glass of Chianti. It's a fabulous dinner—a meal fit for a king!

Chestnuts and Vino Nuovo

October is also the season for chestnuts, and since the fire is already lit for the mushrooms and the meat, we can roast a few chestnuts for a wonderful end to the evening.

The hill where one of our vineyards is now located is called Chestnut Hill because there used to be a lot of chestnut trees there.

Today there is only one lone chestnut tree left on this hill. It grows right at the top, surrounded by vines, but because it's a wild chestnut tree, its nuts are inedible. So my aunt Maria Moschi brings us chestnuts from her little property in San Gimignano.

The special iron pan for cooking chestnuts has holes in the bottom and a long handle. After you've scored the chestnuts, meaning you cut a small slit in the shell of each one, you put them in the pan, put the pan directly in the flames, and cook the chestnuts for twenty minutes or so. I can hardly wait for the chestnuts to come out of the pan before I grab some, peel them, and pop them into my mouth. I almost always burn my tongue, but they're so delicious, with their floury texture and sweet flavor, that I don't care. Chestnuts are traditionally served with *vino nuovo* ("new wine"), freshly made wine from the new harvest that's not ready to be bottled yet.

In the tradition of simple Tuscan recipes, there's an excellent dessert made with chestnut flour called *castagnaccio*. At one time, chestnuts were one of the staples of a farmer's diet. *Castagnaccio* is a flat, thin dessert made with chestnut flour, raisins, walnuts, and extra-virgin olive oil. Today it is generally made with pine nuts.

Insalata di Fungi, Parmigiano Reggiano e Rucola

Mushroom Salad with Parmesan Cheese and Arugula

Serves 6

9 ounces (180 g) baby arugula

4 cups (300 g) sliced white mushrooms

2 tablespoons (30 ml) extra-virgin olive oil

Juice of 1 medium lemon

½ teaspoon (3 g) sea salt

Small wedge of Parmesan cheese, preferably Parmigiano-Reggiano

Wash the arugula, drain well, and pat dry using absorbent paper towels. Combine the mushrooms and arugula in a bowl. Drizzle the olive oil over the greens, and add lemon juice and salt. Toss well and divide among individual serving plates. Using a vegetable peeler, shave about 4 very thin slices of the cheese on top of each salad. Serve at once.

FARRO

Farro is an ancient grain that has made a comeback after decades of neglect. A member of the genus *Triticum*, farro is one of the oldest types of cultivated wheat. It is believed to have originated in Palestine and then spread to Egypt and North Africa. Many of the farro recipes we still prepare today date back to the Middle Ages, when it was a frequent ingredient in soups, breads, and desserts. The ancient Romans called farro *triticum* or *farris* and often used the grain in dishes that were offered to the gods in propitiatory rituals. It was also used to treat fever.

Farro is the Italian name for *Triticum dicoccum*, known as emmer wheat in the United States (although it's sometimes referred to as spelt, which is actually a different species, *T. spelta*). Once regarded as "poor man's wheat," farro is highly nutritious, rich in vitamins, minerals, and fiber, but low in fat. Its yield, however, is about six times smaller than that for wheat, which is probably why it was supplanted by more desirable types of wheat. Now it is again being grown in many areas where organic farming is favored because it's very resistant to pests and grows even in a cool climate.

Today farro is an ingredient in a variety of recipes, though it is often paired with legumes, which compensate for its low level of essential amino acids. Farro is clearly a nutritional powerhouse, and it's not too heavy and very tasty and lends itself to many different dishes. It's often used as a substitute for pasta or rice. In addition to whole farro, at our farm and elsewhere you can purchase farro flour and farro pasta. Farro pasta is dark and looks like whole wheat pasta. It goes well with many different sauces but is especially delicious when drizzled with olive oil and sprinkled with grated Parmesan cheese.

Insalatina di Farro

Farro Salad

Serves 6 to 8

2 cups (380 g) raw whole farro

10 cups (2.5 L) cold water

1 tablespoon (20 g) sea salt

1 (16-ounce) (473 ml) jar giardiniera (pickled vegetables), drained and cut into ½-inch (1.5 cm) dice

1 large ripe tomato, cut into ½-inch (1.5 cm) dice

1 (12-ounce) (340 g) jar artichoke hearts, drained and quartered

2 celery stalks from the middle of the head, chopped, plus about 2 tablespoons (9 g) chopped celery leaves

2 carrots, cut into ½-inch (1.5 cm) dice

Sea salt to taste

3 tablespoons (45 ml) extra virgin olive oil

Place the farro in a large bowl and add water to cover by 2 inches (5 cm). Cover with plastic wrap and set aside to soak overnight, or for at least 8 hours. Drain well.

CS

Combine the farro, cold water, and sea salt in a heavy-bottomed, 6-quart (6 L) saucepan over medium heat. Cook for about 20 minutes, or until farro is al dente. Drain in a large fine-meshed strainer. Run under cold water until the farro is cooled. Shake out excess water, then transfer to a large serving bowl.

CS

Add the vegetables and toss to blend well. Season with salt. Add the olive oil and stir to moisten all ingredients. Serve at room temperature or chilled.

Zuppa Ribollita

Ribollita Soup

The traditional ingredient in this soup is lacinato kale—cavolo nero (meaning "black cabbage") in Italian. Any type of kale may be substituted.

Serves 6

4 (15-ounce) (425 g) cans cannellini beans

1 cup (240 ml) extra-virgin olive oil

⅔ cup (100 g) chopped red onion

2 celery stalks, cut into ½-inch (1.5 cm) dice

2 carrots, cut into ½-inch (1.5 cm) dice

1 (10.75-ounce) (310 g) can tomato puree

⅓ cup (80 ml) plus 3 cups (720 ml) hot water

7 ounces (200 g) lacinato kale

8 ounces (225 g) Tuscan bread, sliced ½ inch (1.5 cm) thick and toasted

Puree the beans in blender until very smooth; set aside.

ෆ

Heat the olive oil in a heavy-bottomed, 6-quart (6 L) saucepan over medium-high heat. When oil is hot, add the onion and cook, stirring often, until wilted and transparent, about 5 minutes. Add the celery and carrots and cook, stirring often, until the vegetables are lightly browned, about 10 minutes. Add the tomato puree, then rinse the can with ⅓ cup (80 ml) of hot water and add the remaining puree to the pot. Stir to blend. Tear the fibrous midrib from the kale, then tear the leaves into bite-size pieces and add to the soup. Simmer for about 10 minutes. Add the pureed beans and stir to mix well. Add the remaining 3 cups (720 ml) hot water and simmer for an additional 30 minutes.

ෆ

Remove 2 cups (480 ml) of the soup and set aside. Add the bread slices to the pot and cook at a brisk simmer for 15 minutes. Remove from the heat and stir to blend the bread well. Set the soup aside for about 15 minutes, then check the consistency. If it is too thick, add all or a portion of the reserved soup. Reheat, if needed, and serve hot.

Amico harvesting black cabbage.

Penne alla Boscaiola

Lumberjack Mushroom Penne

Serves about 8

½ cup (50 g) red onion, cut into small dice

4 tablespoons (60 ml) extra-virgin olive oil

3 cups (300 g) fresh or frozen white mushrooms,
cut in small pieces

10 ounces (285 g) ground pork sausage

3 cups (680 g) tomato puree

½ cup (120 ml) warm water

½ to 1 teaspoon (1 to 2 g) crushed red pepper, or to
taste

3 pinches salt, or to taste

4 to 6 quarts (4 to 6 L) water

2 tablespoons (40 g) salt

1 pound (484 g) penne pasta

½ cup (45 g) grated Parmesan cheese (optional)

Cook the onion with the olive oil in a medium, heavy-bottomed pot for about 5 minutes. Add the mushrooms and cook an additional 10 minutes. Use your hands to break the sausage into smaller chunks and add to the pot. Cook for 10 to 15 minutes or until the sausage is cooked. Add the tomato puree, warm water, red pepper, and salt. Simmer for 30 minutes over medium-low heat.

In another large pot, bring water to a rolling boil. When the water is boiling, add salt and the penne pasta. Cook pasta until al dente, about 7 minutes, using the package instructions as a guide. Remove and drain.

Put the sauce in a large frying pan and add the pasta. Stir over medium-high heat for 3 minutes and serve. Top each serving with Parmesan cheese, if desired.

Fettine ai Funghi

Sliced Beef with Mushrooms

Serves 8

¾ cup (180 ml) plus 3 tablespoons (45 ml) extra-virgin olive oil

2 cups (150 g) sliced crimini or porcini mushrooms

2 cups (150 g) sliced white mushrooms

3 garlic cloves, minced

½ cup (20 g) minced flat-leaf parsley leaves

¾ teaspoon (3 g) sea salt

⅓ cup (80 ml) hot water

2½ pounds (2.5 kg) beef tenderloin, trimmed and sliced ½ inch (1.5 cm) thick

Sea salt and freshly ground black pepper

1 cup (150 g) all-purpose flour

3 tablespoons (45 g) unsalted butter

Heat the ¾ cup olive oil in a heavy-bottomed, 12-inch (32 cm) skillet over medium heat. When oil is hot, add the mushrooms, garlic, parsley, and salt. Cook, stirring, for about 5 minutes. Add the hot water and cook over medium-low heat, stirring often, until liquid has evaporated, about 15 minutes. Set aside.

ଓଷ

Use a meat pounder to flatten the tenderloin slices to a thickness of about ⅜ inch (1 cm). Season both sides of the meat with sea salt, then season only one side of each slice with black pepper. Place the flour on a plate and dredge the meat slices to coat both sides; shake off excess flour. Set aside.

ଓଷ

Combine the remaining 3 tablespoons (45 ml) olive oil and the butter in a heavy-bottomed, 14-inch (35 cm) skillet over medium heat. When butter is melted, add the meat and sauté for 4 minutes for medium rare. Turn the meat, then spoon the reserved mushrooms and any pan juices over the meat. Cook for an additional 3 minutes on the second side, or longer, depending on the preferred doneness. Serve the sliced beef topped with mushrooms.

CHIANTI

Chianti is a well-known red wine produced in Tuscany. The name Chianti may derive from the Etruscan word *clante,* meaning "water," or possibly from the Latin word *clangor* ("noise"), in reference to the noise made in the region's woods while boars were being hunted. Chianti was also once the name of a region established by the government of Florence. Known as Lega del Chianti (League of Chianti), it included the towns of Gaiole, Radda, and Catellina. It's unclear whether the wine was named for the region or vice versa.

One of the world's most renowned types of Chianti is Chianti Classico, which is produced in an area that still includes Gaiole, Radda, and Castellina but also Greve and a part of Barberino Val d'Elsa, Castelnuovo Berardenga, Poggibonsi, San Casciano Val di Pesa, and Tavarnelle Val di Pesa. In 1932 the Chianti Classico area was better defined, and many of these towns altered their names by adding the phrase "in Chianti"—thus today we have Gaiole in Chianti, Radda in Chianti, Castellina in Chianti, and Greve in Chianti. The Chianti produced in the San Gimignano area is Chianti Colli Senesi.

Chianti Classico is symbolized by a black rooster. According to legend, in the early eighth century Siena and Florence sent out two knights on their fastest horses to establish their borders. At cockcrow, one was to leave from Florence and the other

from Siena, and the idea was that all the land they rode over before they met up would belong to their respective cities. To give their knight a head start, the Florentines selected a black rooster that had not been fed so that it would crow before dawn and give the signal for the knight to depart early, allowing him to cover more ground. The black rooster was also the symbol of the Lega del Chianti, created in 1384.

Chianti has a very long history. It was made purely of sangiovese grapes until the 1700s, when sangiovese began to be combined with other varietals such as canaiolo, colorino, and black malvasia. Later, green grapes such as trebbiano were added as well. The famous recipe for Chianti devised by Prime Minister Bettino Ricasoli in the 1800s called for a blend of 70 percent sangiovese,

20 percent canaiolo, and 10 percent malvasia. This wine became extremely popular and was exported all over the world in those familiar glass bottles covered in straw, which became the hallmark of Chianti.

The main ingredient in Chianti wine remains the sangiovese grape. This red grape varietal has Etruscan origins and has always been one of the most widely cultivated in Tuscany. The sangiovese grape is ruby red in color, and its aroma calls to mind red fruits and flowers such as roses and violets. It has a substantial amount of tannin, and there are also many "clones" of this grape. Other historic Tuscan varietals include canaiolo, with its lovely aroma, sweet colorino, and black malvasia, which imparts a great deal of color.

An old winemaking method for Chianti called *governo alla Toscana* consisted of drying 10 percent of the grapes on mats. Then they were pressed, and the resulting must was added to the new wine before the month of December, or sometimes in the spring, in a process known as *rigoverno*.

Chianti was considered a good wine until the early 1900s. Around that time, Italy's sharecropping system resulted in a focus on quantity over quality when it came to grapes, and the product declined. Today Chianti is once again considered an excellent wine, and it is still one of the most famous wines in the world. Chianti is fairly dry and lightly tannic. It pairs well with grilled red meat, stews, and braises, and it goes quite nicely with various cheeses as well.

NOVEMBER

The weather begins to turn cold. The temperature drops and the days grow shorter. Hunting season starts in mid-September, but the kind of hunting that draws the most attention begins at this time of year—boar hunting. Every boar hunter has his own hunting license and goes out with his own group. Specific hunting rules, not to mention moral guidelines, must be followed.

Often we see dozens of people parking their cars on both sides of the road, heading off to hunt. They tromp into the woods, rain or shine. They're all wearing camouflage and have rifles on their shoulders. They also have their hunting dogs on leashes. All day long we hear shots being fired in the woods, followed by people whistling to call their dogs. Some groups of hunters are lucky and have a fruitful hunt, while others return empty-handed. After sundown they aren't permitted to hunt any

longer, so all of them—the lucky and the less so—head home.

Fortunately, my cousin Giancarlo and my uncle Piero Moschi are in two different hunting groups, so we always manage to get boar meat from one or the other. Often on Friday we eat homemade tagliatelle with boar sauce as well as boar prepared *alla cacciatore.* The meat is cooked with bay leaf, black olives, and red wine and then tomatoes are added. The bay leaf and the red wine temper the gamy taste of the boar meat.

Night is falling earlier now as the days grow shorter. We use a little tractor-pulled cart to transport all the pot plants and potted lemon trees to the fruit cellar before it gets really cold. I spend the day plucking flower pots from the windowsills, the terraces, and the garden. They will all be kept in a warm place to protect them from the winter frosts

and the cold temperatures that are on their way.

Harvesting and Cleaning Saffron

The frenetic pace of our busy tourist season slows somewhat, although there's plenty of work to do in the fields. Between late October and early November, small saffron flowers begin to send up their blossoms. These lilac-colored flowers really light up the subdued autumn fields. In this season the vineyards are dark red and yellow, and the fields are brown. Delicate-looking saffron flowers offer a welcome contrast. Saffron is planted in August and blooms in the fall for about two weeks.

When planted, the bulbs must be spaced at least four inches (10 cm) apart, in a plot at least thirty-nine inches (1 m) wide; the length doesn't matter. Two small drainage ditches must be dug around the sides of this plot so

that water doesn't collect and stagnate. It's advisable to protect saffron with netting as it grows because boars and porcupines can uproot the bulbs. In December and January we check the crop regularly; any bulbs that appear diseased are dug up and destroyed to prevent the spread of a fungus that could infest the entire crop.

Bulbs are kept a certain distance from each other and dug up at the end of their vegetative period, stored, sorted, and then the healthy ones replanted. The bulbs reproduce annually by forming new corms.

We have found that using the same bulbs and the same land for a three-year cycle results in the following pattern. The first year the yield is minimal. Yield peaks in the second year, and then in the third year it falls again, though it's still larger than the first year. Saffron must be harvested carefully by hand,

SAFFRON

Saffron—as indicated by its Latin name, *Crocus sativus*—is a member of the crocus family. In Greek myth the gods opposed the love between Crocus and the nymph Smilace, so they punished Crocus by turning him into the flower that goes by his name. According to Roman myth, however, Mercury threw his discus poorly and hit his friend Crocus by mistake, so he tinted the flower with his friend's blood and named it in his memory.

Saffron has ancient origins that can be traced to an area extending from the Middle East to Crete and Greece; it was soon widespread in the Mediterranean and northern India. Frescoes painted in Knossos as early as 1600 B.C. depict butterflies plucking saffron flowers. After the death of Buddha in 480 B.C., saffron became the official dye for coloring the clothing of Buddhist monks. The saffron trade to the West was well developed early on.

In medieval times the sale and production of saffron in San Gimignano flourished because the Via Francigena passed through the city, which grew rapidly as a result. Several documents in the San Gimignano archives mention saffron. In 1228, for example, the city paid the debts stemming from the attack on Castello della Nera in both money and saffron. In 1276 tax duties were levied on the import and export of this spice, and in 1295 San Gimignano assigned two officials to stand at the city gates to weigh any saffron passing through and accept the fees related to transporting it. Saffron was grown in several places around San Gimignano, such as the old village of Casaglia, which lies a short distance down the road from Poggio Alloro.

Saffron was traditionally used in a wide variety of ways: as a textile dye, a

medicine, a culinary spice, and even a form of currency. Today it is used only for cooking, and it is produced in several parts of Italy, including Umbria, Abruzzo, and Sardinia. Saffron of varying quality is also grown in North Africa and the Middle East.

This plant does not produce any seeds (the bulbs sprout new bulbs), so it must be cultivated. Enclosed in each flower's six petals are three stigmas, which are the source of the spice that has been used by cooks for thousands of years and in many different ways. Removing these delicate filaments from the harvested flowers is an extremely labor-intensive task. In Middle Eastern countries this is often women's work, as their smaller hands and fingers make it easier.

Once the saffron has dried for a couple of days, about 85 percent of its moisture evaporates, leaving it much lighter. It takes about 350 flowers to yield just one gram (0.035 ounce) of saffron stigmas. Perhaps saffron is not described as golden yellow just because of the color it lends to dishes, but also because its market price is higher than that for unrefined gold! At this writing, the average market price is 35 euros per gram. Yes, you read that correctly! The high price is due to the intensive amount of work required for the entire process. Everything from tilling the soil through planting through harvest and the cleaning of the flowers is done by hand; it simply cannot be done by a machine.

If you're looking for quality, obviously it's better to purchase whole saffron stigmas rather than the powdered form. Sometimes the powder includes other spices, such as paprika. The stigmas must be soaked in liquid—water, broth, or milk, depending on the dish—in order to release the scent and flavor of the spice.

preferably early in the morning or late in the evening when the flower petals are closed. This makes it easier to pick the saffron flowers and to work more quickly; when the flower is open, you risk damaging the saffron stigmas, as they are often touching the petals and may stick out of the flowers slightly. Once saffron flowers have been picked, they are placed in wicker baskets and brought back to the farmhouse. There the entire family sits by the fireplace, opens each flower, and pulls out its three red stigmas, which are then placed in sieves to dry near the fireplace (or in the sun if it's a sunny day).

Our 2008 saffron crop was particularly large, which meant that we often spent the entire day cleaning saffron flowers, starting early in the morning and finishing at dinnertime. My parents and my aunts even continued to work at night. They'd bring a basket of flowers with them when they sat down to watch television after dinner. We had so much work that we enlisted not just family members but also all our employees. So in addition to my parents and my aunts Maria, Giannina, and Maria Moschi, there were Vilma, Mimosa, Tosca, and Anna working away. Billo would watch us from his cozy spot next to the hearth, while Cila, Billo's mother, preferred to sit on her own seat next to Aunt Moschi. We all had pieces of yellow paper in front of us

and a bunch of flowers that had already been opened. After a few hours of work, someone would always realize that he or she was tossing the saffron into the flower pile and shout, "Oh, no!" I have to be honest here: this was much more likely to happen to the older members of the family than to the younger ones.

We also got into the habit of having tea and *sfogliatine* pastries mid-afternoon. The place that makes the *cantuccini* cookies that we sell at the farm also makes excellent *sfogliatine*. I would prepare either tea or coffee, and we'd take a break and have a snack. While we sipped tea, people would start telling stories about what had happened that season. One time it was Vilma's turn to be teased. We reminded her about the time she gave an American guest a key to room number eight and then said, in English, "Room number . . . eight o'clock!"

Then there was the first time we had an English-speaking group for lunch, and they asked, "Where is the restroom?"

Vilma said, "Why are they asking where the restaurant is? The *ristorante* is here; they're standing in it!" We had a good laugh, and then we focused once again on our work.

I remember one evening when we were exhausted after having cleaned thousands and thousands of flowers, but we were so satisfied with ourselves. Just as we were getting up

Above: Maria Moschi, Umberto, Rosa, Maria, and Amico.
Right: Amico and Rosa.

from our chairs, Amico entered. I said, "Look, Babbo, look what a good job we did. We cleaned all those flowers!"

And he said, "Yes, very good job. Now get to work on these," and he emptied the contents of another two baskets of saffron flowers onto the table.

That year we hardly had time to pick the flowers before another hundred or so bloomed—it was really quite incredible to see. The morning after the harvest the field would be completely green, and as little as a half hour later it was filled with waves of lilac-colored flowers.

Sowing

Fields are plowed and then in November they're ready to be seeded with hard and soft wheat, farro, barley, and oats. During medieval times grains like wheat, farro, oats, barley, and millet were consumed in large amounts. Legumes were also an important food, and the most commonly cultivated were fava beans, chick peas, peas, and beans.

Hard wheat is used to make semolina flour, which is used for pasta, while soft wheat has more starch and is used to make flour for bread, cakes, cookies, and other desserts as well as pizza. Often both kinds of flour are combined in homemade egg pasta. Our pasta is made simply with water and semolina flour only. It's shaped using bronze molds and dried naturally. That's why our pasta has such a natural color and the true flavor of grain.

One day, when Amico was on the Caterpillar tractor and plowing a field to prepare it for planting, the plow suddenly hit something hard in the ground—lumps of brass and iron. Upon closer examination, he saw that they were the exploded rounds of a 120 gun (120 mm in circumference) from World War II. When he first saw the ammunition, he jumped off the tractor and ran away in fear. Once he'd established that the situation wasn't dangerous, he returned to the area and found a total of eight rounds. Once, when butchering pigs, he even found a pair of machine gun bullets in their stomachs. He tells me that they "shot like crazy" around here.

Our land has some other surprises under the soil, in addition to souvenirs of the war—thousands of fossils of all kinds. After a heavy rain, we can see many of them in the bare areas where there's no grass. Millions of years ago, this area was actually under the sea.

My father still brings me the prettiest fossils that he finds in the vineyard.

The particular make-up of our soil, which is very rich in minerals, gives the vernaccia wine in our area its signature salinity.

Harvesting Olives

On our farm we once had olive trees that were centuries old. The most recent of our three record freezes, in 1985 (the other two were in 1929 and 1956), killed the remaining olive trees, and the older trees were struck down in most other areas, too. That year, all the olives dried up, so no oil could be made. Amico still gets a little teary-eyed when he talks about those days. The temperature dropped precipitously to -8 degrees F (-22 degrees C). The oldest trees—some as much as seven hundred years old—were the ones that suffered the most. After that, we had to wait five or six years for new trees to grow and mature enough to produce olives. Poggio Alloro has four varieties of olive trees: Frantoio (or Frantoiano), Moraiolo, Leccino, and Pendolino.

As my father always says, though, nature finds a way to help you out. Some small shoots in the earth that were covered with snow survived the freezing temperatures and grew around the trunks of the old trees that had died off. That's why many of the olive trees on our property appear to have three or four trunks rather than just one—if you look in the middle, you can see the trunk of the old tree that was cut down. The ancient olive trees were beautiful, and some were so large that it would have taken more than two people to reach around their trunks. It's truly a shame that they're gone from the farm.

The secret to the long lives of our olive trees is the fact that the olive tree is perfectly suited to the Mediterranean climate, with its mild winters and hot, dry summers. Another key factor in the long life of our olive trees is the farm's altitude above sea level. In particularly hot areas, the olive tree grows best at 2,300 feet (700 m) above sea level in order to mitigate the effects of the warm climate. After an olive tree flowers in May and June, there is a setting period when the fruit forms. Only a small portion of the flowers will set. Most olive trees are fertilized via pollination through the wind. In the subsequent months, the fruit grows, yet even olive trees that seem to be growing a large crop of olives may not produce much of a harvest. If the summer is very hot, olives will fall from the trees early, and the harvest in November will be much scantier than anticipated.

Throughout September the olives ripen and turn from green to dark purple. At the

Clockwise from top left: Fermo, Vico, and Bernardo in the olive trees. Fabio. Bernardo. Mario (left) and Moreno (right). Fermo. Bernardo and Paolo.

OLIVE OIL

Olives are among the oldest crops in the Mediterranean. In Rome and Greece in the second century B.C., olive oil was already a key source of fat in the everyday diet, and when not enough of the oil was produced, it was imported from Spain. Romans learned from the Etruscans to use olive oil in their cooking.

Olive oil has traditionally had many uses. Pharmacies and abbeys distributed it for medicinal purposes, and it was mixed with wine and egg whites to create "the Samaritan's balsam," a salve for burns. Athletes spread the oil on their bodies before competitions to counteract sudden changes in temperature and increase muscle tone. An Egyptian scroll discusses its use as an ointment and an ingredient of creams and pomades. Romans used it to whiten their teeth, and today a mixture of olive oil and avocado oil is used for tanning. The oil was considered a purifying element, a symbol of life and hope. It was burned in lamps and was part of the liturgy of various religions.

Olive cultivation dropped precipitously when the Roman Empire fell and the Barbarians invaded, as they had a completely different diet and preferred to use land for raising livestock rather than growing olives, which was an increasingly difficult and expensive activity. Soon olives were grown only inside the walls of convents. The olive had a bit of a renaissance during the era of the medieval communes and was popular in Europe in the 1700s.

Olive oil is still considered part of a healthy, balanced diet and preferable to margarine, butter, and other fats. Our grandparents used the oil for medicinal purposes, dosing it out by the spoonful to children with stomachaches. Women still use it today to soothe chafed nipples when nursing, and it is commonly found in all kinds of cosmetic preparations. Over the last twenty years, the oil has had periods of widespread popularity as a dietary fat, alternating with periods when it was shunted aside in favor of butter and oils made from seeds, which are considered easier to digest. Olive oil works with any kind of food, although it's not suitable for frying because it cannot hold high temperatures for long

periods (i.e., it has a low smoke point).

In Italy olive oil is produced in several regions in addition to Tuscany, mainly Sicily, Puglia, Umbria, and Liguria. Each area has its own cultivars of olive trees. In Tuscany the most common are Frantoio (Frantoiano),

Moraiolo, Leccino, and Pendolino.

When olives are taken to the mill, they are first cleaned, a process that removes greenery and washes the olives. The oil is then extracted using the pressure of the presses. Next the oil and water (oil must) are separated from

the solid olive residue of pits and pulp (sansa). All this is done under cold conditions. If machinery is allowed to heat up, it can result in a larger yield of oil, but it will also change the oil's organoleptic qualities. Finally, the oil and water are separated from each other, and after being decanted briefly the oil is ready for use. Yield in the San Gimignano area hovers between 13 and 18 percent—the percentage of oil obtained from every 100 kilograms (about 220 pounds) of olives.

Olive oil is best left unfiltered and stored in the right kind of container. The oil used to be kept in a terra cotta amphora, but today it is typically stored in a dark green bottle to protect it from light. It should also be kept at a constant temperature and away from heat. For a month after it is pressed, the oil is opaque, but after that it naturally clarifies and turns a clear golden

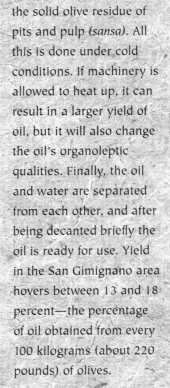

green. Most oil can be stored for fourteen to eighteen months.

To be labeled "extra-virgin" (the highest grade of olive oil), the oil must come from a first pressing and have an acidity of less than 0.8 percent. The less acidity, the better the oil (Poggio Alloro's olive oil has an acidity of about 0.2 percent). Oil that is unfiltered and organic is better, but these attributes don't affect the extra-virgin designation. The categories for olive oil established for European producers are as follows:

First pressing:
extra-virgin olive oil

Second pressing:
virgin olive oil

Third pressing:
olive oil

Fourth pressing:
olive pomace oil

Tuscan olive oil is usually greenish in color and smells fruity and grassy, with a hint of bitterness and spice, especially when it is newly pressed.

same time, they become drier as their water content decreases. In our part of the world olives are fully ripe in November.

Depending on the season, the trees are pruned between February and April, when the olives aren't wet. Traditionally pruning never takes place on the last three days of January, which are known as the "blackbird days" (*giorni della merla*). These are the coldest days of the year, and according to superstition, if those three days are really cold, spring will be warm; if the three days are warm, the spring will be a cold one. According to legend, a female blackbird with white feathers hid herself and her chicks in a chimneypot to stay out of the January cold. She came out only when the cold had subsided, on February 1, but when she emerged she was so dirty with soot that she was black. And since then blackbirds have been black.

Pruning, performed between the end of the harvest and the beginning of the vegetative stage, is intended to encourage photosynthesis, give the foliage the right shape for that specific crop, and increase production overall. After the branches are pruned, they're picked up off the ground, arranged around the borders of the plots, and then burned to prevent diseases, such as those from cochineals, from spreading among the olive trees. Plenty of other things can damage the yield of the harvest, however—spring freezes, high rainfall, a lack of rain, and certain parasites, such as the olive fruit fly.

Olives are harvested throughout the month of November. The method that is most commonly used in our area, and the one we use on the farm, is to harvest by hand. Sometimes we also use small plastic rakes to harvest the olives. First, we spread a piece of cloth under a tree and then women stand around it to pick the low-growing olives while the men climb up wooden ladders to pick the olives on the upper branches. The olives are dropped onto the cloth. When a tree has been cleaned completely of olives, the cloth is gathered up on all sides so that the olives roll to the center, and then the olives are dumped into a plastic crate. The crates full of olives will be taken to the mill that evening. It's not wise to put off pressing olives for even a single day, as mold can form due to humidity and the resulting olive oil will be of lesser quality.

The oil was traditionally extracted using a stone press, but almost everywhere today it's processed mechanically by stainless steel presses. Very few farms have their own olive presses nowadays, so farms make scheduled trips to take their olives to a local mill to be pressed. During the peak harvest period, if the mill is fully booked, you may have to deliver your olives in the middle of the night!

Fettunta

Traditional Tuscan-Style Bruschetta

1 slice Tuscan bread per serving, about ½ inch (1.5 cm) thick

1 large garlic clove, peeled and sliced in half, plus additional as needed

Sea salt

Extra-virgin olive oil

Grill the bread slices over a fire or under the broiler. When bread is toasty and crisp on both sides, rub one side of each slice with the cut side of the garlic clove, then season to taste with salt. Drizzle a little olive oil over the bread and serve warm.

Risotto con Zucca Gialla e Zafferano
Autumn Risotto

Serves about 8

4 tablespoons (60 ml) extra-virgin olive oil

½ cup (75 g) red onion, diced

3 cups (450 g) pumpkin, peeled and cubed

Sea salt and black pepper to taste

1 teaspoon (1 g) minced fresh rosemary

1 quart (1 L) hot water

1 quart (1 L) vegetable broth

4 cups (762 g) Arborio rice

½ teaspoon (0.1 g) saffron

1 tablespoon (15 g) unsalted butter

Grated Parmesan cheese (optional)

Heat the olive oil in a heavy-bottomed, 14-inch (35 cm) sauté pan over medium heat. When oil is hot, add the onion and cook, stirring occasionally, until onion is wilted and transparent, about 5 minutes. Add the pumpkin and season with salt and pepper. cook for 20 minutes. If the pumpkin starts to stick to the pot, add some of the hot water Add the rosemary and sauté until the pumpkin begins to soften.

◌◌

While the pumpkin mixture is cooking, combine 1 cup (240 ml) of the water with the vegetable broth and saffron in a heavy-bottomed, 3-quart (3 L) saucepan; cook over medium heat about 10 minutes. Set aside and keep warm.

◌◌

Add the rice to the pumpkin mixture and sauté for 2 or 3 minutes. Add one-third of the vegetable broth mixture, reduce heat to medium, and cook, stirring, until rice absorbs the broth. Add 1 cup (240 ml) of the water, and cook, stirring, until it is absorbed. Continue to add broth and water alternately, cooking and stirring, allowing the rice to absorb the liquid between each addition. When the rice becomes creamy, add the broth and water more sparingly, or the rice will become mushy. Cook until rice is creamy but still al dente, about 25 to 30 minutes total. Stir in the butter and scatter a little Parmesan cheese over the top, if desired, but don't add so much cheese that it overpowers the delicious taste of the saffron.

Cinghiale alla Cacciatora

Braised Wild Boar

Serves 6 to 8

⅓ cup (80 ml) extra-virgin olive oil

2.2 pounds (1 kg) wild boar loin (or substitute pork
  loin), trimmed and cubed

1 cup (230 ml) dry red wine, such as Chianti

Sea salt and freshly ground black pepper to taste

3 garlic cloves, peeled and minced

2 fresh rosemary sprigs, leaves removed and
  minced

1 onion, roughly chopped

3 celery stalks, roughly chopped

4 bay leaves

1 cup (120 ml) water

1½ cups (340 g) tomato sauce

¼ cup Leccino olives, or substitute Kalamata olives

¼ teaspoon (0.5 g) crushed red pepper, or to taste

1 (16-ounce) (454 g) package tagliatelle or
  pappardelle pasta, prepared according to package
  instructions

Heat the olive oil in a heavy-bottomed, 5-quart
(5 L) braising pan over medium-high heat. When
the oil is hot, add the meat and cook until browned
on all sides, stirring often, about 10 to 12 minutes.
Add the wine and cook until the wine evaporates,
scraping bottom of pan to release all of the browned
bits, about 10 minutes. Add salt and pepper.

℅

Add the garlic, rosemary, onion, celery, and bay
leaves and cook for 30 minutes. Add ½ cup (120 ml)
of the water, lower heat, cover, and simmer slowly
without stirring for 30 minutes. Add the remaining
water little by little.

℅

Add the tomato sauce, olives, and red pepper,
stirring to blend well. Adjust seasonings and
cook over low heat, covered, for an additional 30
minutes. Serve immediately over the hot pasta.

Sugo di Cinghiale
Wild Boar Sauce

Serves 6 to 8

1 cup (240 ml) extra-virgin olive oil

2 carrots, diced

2 celery stalks, diced

1 medium red onion, diced

3 garlic cloves, peeled and minced

2 tablespoons (7 g) minced parsley

2.2 pounds (1 kg) ground wild boar

¼ cup (60 ml) Chianti red wine

6 cups (1,360 g) tomato sauce, preferably made
   from San Marzano tomatoes

¼ teaspoon (0.5 g) crushed red pepper

2½ teaspoons (17 g) sea salt

1 (16-ounce) (454 g) package tagliatelle or
   fettuccine, cooked according to package
   instructions

Heat the olive oil in a heavy-bottomed, 6-quart (6 L) saucepan over medium heat. Add the carrots, celery, onion, garlic, and parsley and cook until lightly browned, about 15 minutes. Add the meat and use the back of a large spoon to break up any clumps (the meat should be in small particles). Cook until lightly browned, about 15 minutes. Add the red wine, tomato sauce, crushed red pepper, and salt; cook an additional 25 minutes, stirring occasionally to prevent sticking. To serve, spoon over the hot pasta.

Tacchino allo Zafferano

Golden Saffron Turkey Breast

Serves about 6

2 pounds (1 kg) turkey breast

Sea salt and black pepper to taste

1½ cup (350 ml) extra-virgin olive oil

2 carrots, diced

2 ribs of celery, diced

2 small red onions, diced

½ teaspoon (0.1 g) saffron

Tie the turkey breast (see tying instructions for Vitello Arrosto recipe). Sprinkle with salt and pepper, then place in a large cold pot. Add the olive oil, carrots, celery, and onions and mix well.

CB

Cover the pot and cook over medium heat, stirring occasionally, about 30 minutes, or until the vegetables start to soften. Cut into the meat a little and check for doneness. When the turkey breast is well done, transfer to a cutting board and allow it to rest for at least 10 minutes.

CB

Using a slotted spoon, remove the cooked vegetable from the pot, reserving the olive oil in the pot. Purée the vegetables and transfer to a smaller pot. Add the saffron and cook on low heat for 15 minutes.

CB

When ready to serve, slice the turkey breast ¼ inch (6 mm) thick, transfer the vegetable puree to a serving dish, and arrange a layer of the meat slices over the puree. Top with a couple of spoonfuls of the reserved olive oil.

in cellophane and decorated with brightly colored ribbons and bows. We also prepare a lot of boxes containing two or three bottles of wine for companies in the area, which like to give them as gifts to their employees and their biggest customers. This is also the season for sending Christmas cards. With all this activity going on, the store looks like the North Pole workshop with many elves hard at work.

At the same time, butchering is taking place—this week we'll butcher a Chianina calf and two pigs. Customers stream in and out to pick up the meat, which they take home and cook for their families. Indeed, we refer to these packages of meat as "family packages." They're big white cardboard boxes, each one printed with our name and other information and a picture of a Chianina cow. Inside each package are various cuts of veal, including steaks, stew meat, lean meat, veal shanks, meat for boiling, filets, and more. Each cut is carefully wrapped in vacuum packaging and ready to be cooked immediately or refrigerated. During the week leading up to Christmas, I give something we make on the farm, such as pasta, to each customer who buys something in our store, along with a holiday card.

Our holiday cards are always personalized. For a few years I used a photo of all the members of our family together; another series showed the three brothers making a toast. This year I used a photo of two of our Chianina cattle grazing with Santa Claus hats on their heads.

Christmas is a special time, and I've always loved it. I like to decorate every room in the farmhouse, and I put Christmas trees everywhere I can. The tree in my parents' house, obviously, is the one I dedicate myself to the most. I still use the balls and other ornaments we had when we were little. Decorating the Christmas tree was the responsibility of the oldest sister, and then the middle sister and then, finally, after I'd waited years and years, the task was entrusted to me.

In Italy it's traditional to set a crèche (*presepe*) under the tree to depict the birth of baby Jesus. A crèche is arranged by spreading a layer of green moss under the tree to make a small field. We gather the moss in the woods a few days before Christmas, and it stays fresh until January 6. Then we create a path of pebbles or sand and then place small figurines of Joseph, Mary, the ox, the ass, and the manger with baby Jesus inside a small lean-to. Around that go small houses, grazing sheep, shepherds, and village women standing around a small lake with geese skimming across the surface. Three camels and the Three Wise Men are in place, too, and they move closer each day, arriving in time for the Epiphany. Finally, a shooting star rests on the roof of the lean-to and glows in the dark. Or rather, it used to glow in the dark; it doesn't light up anymore. Like all the pieces of our crèche, it's quite old and has been passed down for generations. In fact, one of the shepherd figurines has to be positioned with his back to the wall so that the back of his head isn't visible—one year Billo decided that this poor shepherd was the best toy ever. Before we could stop him, he had picked it up in his mouth and managed to chew up the head and bite off one of the hands. At Christmas every home and every church has its own crèche. Some towns even have living crèches, with people in costume acting out the nativity scene.

In my family, as in many other families, Christmas Eve is a major holiday. We place our gifts under the tree and keep them there in the days leading up to Christmas so that we can open them all together on Christmas Eve. Families with children often receive visits from Babbo Natale, or Santa Claus, in the form of a neighbor who dresses up and goes from house to house to bring gifts.

Christmas is the only holiday when the farm is closed, so it's the one day of the year when we can all be together. On Christmas Eve we adhere to tradition and eat fish. Amico goes to the market in the piazza each year and buys fresh fish. We don't normally cook fish on the farm, because we produce beef and as a rule we serve only products from the farm. Luckily, though, my mother grew up in San Benedetto del Tronto, which is on the Adriatic coast in the Marche region, and she's very skilled at cooking seafood. It takes my mother two days to clean and prepare all the fish.

We eat a light lunch on December 24, something like canned tuna and vegetables. That night we have a real feast on a table decorated with Christmas ornaments and a candle in the center. We always enjoy a special wine in addition to our vernaccia, which matches beautifully with fish. Our meal starts with a variety of fish-based antipasti. The pasta course is everyone's favorite—linguine with seafood, such as mussels, clams, cuttlefish, and shrimp, in a light tomato sauce.

By the time we get to the second course, usually swordfish, tuna, and dorade with roasted potatoes and tomatoes halved and topped with breadcrumbs, we're all pretty much full already. We always have salad, and then we finish off the meal with typical Christmas desserts such as *torrone,* bars of nougat, and *pandoro* and panettone cakes. Then we go right on eating, snacking on dried fruit and walnuts, hazelnuts, almonds, and more.

When we're truly full and ready to unwrap our gifts, we gather to sit on the floor around the tree. We take packages from under the tree and pass them around. Some of us like to shake them and try to guess what's inside. After each gift is opened, there are thank-yous and kisses. Our celebration goes on late into

Right: Rosa, Amico, Maria Moschi, and Giannina make tortellini as Gena looks on.
Lower right: Laura, Filippo, Sara, and Sarah.
Below, center: Vilma and Marco. Bottom: Vanessa with the Christmas pandoro.

the night. When it's late enough that we know everyone has finished eating, we usually visit each other's homes to wish each other happy holidays.

Usually my extended family gathers at my house. The younger cousins and my little nephew play a card game called seven and a half *(sette e mezzo)*. They bet anywhere from ten to fifty cents on the games. Young and old sit together and talk and laugh. We describe the gifts we've received. We occasionally shell hazelnuts and snack on a few, though it's impossible for anyone to be hungry after all we've eaten. Before sitting down to play cards, Rosa asks who wants coffee and then carries in small cups of steaming espresso. The conversation is lazy and friendly: *Who wants sugar? Whose turn is it? Oh, he got my fifty cents! No, it's not your turn. It's his. One and a half spoonfuls of sugar, thanks. Quit eating those hazelnuts! Come on, pick a card! Your turn!* This goes on until midnight, when some in our group bundle up to go out to midnight mass, which is always beautiful to see, and others go to bed, planning to get up early and go to morning mass.

On Christmas morning we have a breakfast of milk and coffee and *pandoro,* while Rosa finishes preparing the tortellini for lunch. She places a bit of stuffing in each small square of pasta, then closes them all so quickly that she looks like a machine. They'll be cooked later in capon broth. (A capon is a rooster that is castrated to make it more tender and also to help it gain more weight.) Amico is sitting on the couch next to Billo. He's pretending to watch TV, but really they're both asleep. One of the two is snoring, too, but I can't tell whether it's the dog or my father!

On Christmas Day, too, the whole family gathers around the table for lunch and dinner and for meals the next day as well.

Amico is usually responsible for making a toast to wish everyone a merry Christmas, and then we all begin eating an antipasto of Tuscan crostini, vegetables preserved in olive oil, and prosciutto and salami.

The first course is hot broth with

TRADITIONAL CHRISTMAS MENU

❧

Chicken liver crostini

Homemade tortellini in capon broth

Capon with cooked greens and green sauce

Roasted chicken and pork

Salad

Traditional Christmas desserts: pandoro, panettone, ricciarelli, torrone

Dried fruit, almonds, walnuts, hazelnuts, pistachios

Coffee and liqueur

tortellini; when it's served, the room grows silent, as everyone is busy enjoying the excellent flavor. We all pay compliments to Rosa. That's followed by meat from the capon that was boiled for several hours to make the broth, along with boiled vegetables and a *salsa verde* of parsley and garlic.

Today for Christmas I've made a special dessert, a *pandoro* cake in the shape of a Christmas tree. *Pandoro* is perfect for this. All you have to do is cut slices and then fan them out slightly for the branches. Each layer is topped with pastry cream or chocolate and then sprinkled with green powdered sugar. It's topped with ornaments made of candy, and in place of the lights I use birthday candles. It's an edible Christmas tree!

Making Vin Santo

One of the most important jobs to be done in December is the making of *vin santo,* a typical Tuscan dessert wine that's very strong. It's been made from the must of dried grapes for hundreds of years. The name *vin santo,* or holy wine, appears to derive from the fact that the wine was once used to celebrate mass. Also, in the past the wine was drawn off during Holy Week, hence the name holy wine.

The grapes for this wine are usually harvested in mid-September, and the best fruit is chosen after it has had a chance to develop a large amount of sugar and is extremely ripe. At our farm we use white malvasia, San Colombano, which has become very rare in Tuscany, and a small percentage of colorino, which gives the *vin santo* sweetness and a beautiful amber color that lasts even after years of aging.

Once the grapes are harvested, they are painstakingly hung, one at a time, in a kind of pyramid. Each bunch of grapes hangs on its own hook. The grapes are left to dry for about two months. Traditionally the grapes were spread out on straw mats called *cannicci,* and then they were turned so that they dried on all sides. However, when the grapes were being turned, some grapes fell off the stems and were ruined, so we developed our own unique system—the bunches of grapes are a certain distance from each other and hang in the air, so they don't need to be touched and they don't touch each other.

While the grapes are drying, the days are hot and dry, and we leave the windows open to get as much air circulation as possible. The bunches of grapes dehydrate slowly, and as the water in them evaporates they grow increasingly sweet. In December the grapes are then pressed, and the must that results is fermented in wooden barrels of various sizes that are filled two-thirds of the way up. If the barrels were filled up to the top, they'd explode during fermentation. A starter culture known as the *madre* is placed in the barrels with them. The *madre* is sediment that collects in the barrels and consists of enzymes and yeast. There are two schools of thought about using the *madre* in *vin santo:* Oenologists believe that it's unnecessary because the wine will ferment without it. Farmers, however, believe that using the *madre* is important, as it gives the wine its unique characteristics. The *madre* is never discarded, but instead is transferred from barrel to barrel. Bernardo still makes *vin santo* the old way.

During fermentation the sugar concentration grows so high that the yeasts are not able to transform all the sugar into alcohol, so our *vin santo* is always about 16 percent alcohol and therefore maintains a high level of sugar. The barrels that are used to make our *vin santo* have been in our family for many generations—some of them are more than one hundred years old. These barrels are never replaced, as *barriques* are, but are used over and over for years. Some of our oldest barrels are made from the wood of cherry and chestnut trees; others are mulberry and durmast oak.

During fermentation and aging, *vin santo* needs to experience wide swings in temperature from day to night and as the seasons change. It is because of these fluctuations that the process takes three years. The wine ferments in the barrels, which are closed and sealed with sealing wax or cement. We arrange our *vin santo* barrels in a row under the eaves. Before our new wine cellar was built in 2001, we kept the barrels in the attic.

After three years, the *vin santo* is drawn off, meaning the liquid part is taken out, the dregs are removed, and then the wine is returned to the barrels. This time they are filled to the brim and sealed again. Then the wine is left to age for an additional three years. *Vin santo* has a very small yield: about 220 pounds (100 kg) of grapes will result in only ten to twenty quarts (10 to 20 L) of wine. Making *vin santo* requires patience and time. Indeed, there are few places left where it is made using the old methods. Several winemakers have begun cutting short the aging periods and are researching other methods for shortening the time it takes to produce this dessert wine.

Traditionally *vin santo* is served with biscotti di Prato, also known as *cantuccini.* These are very dry cookies made with eggs, flour, sugar, honey, and, most important, almonds. Because *cantuccini* are so dry, they are dunked in a glass of *vin santo* and softened and flavored in the wine one bite at a time.

Insalatina di Finocchi
Fennel Salad

About 8 servings

2 large fennel bulbs

2 oranges

1 small can pitted, sliced black olives or
   pitted Kalamata olives

3 tablespoons (45 ml) extra-virgin olive oil

Sea salt to taste

Wash fennel thoroughly and cut in half
lengthwise; then cut each half into thin slices
across. Cut the orange in half lengthwise
and then cut each half into very thin slices
across, leaving peel on. Put the fennel in the
middle of a plate and arrange the orange
slices all around the fennel. Garnish with
1 orange slice and the olives. Drizzle with
the olive oil and sprinkle with salt.

Amico harvesting fennel.

Crostini Neri (Crostini Toscani)

Crostini with Tuscan Pâté

Tuscan crostini, also known as crostini neri *("black crostini"), is one of the signature dishes of Tuscany. The recipe we use on the farm has been modified a lot to make the most appetizing version of the dish for visitors. The original recipe called for chicken livers, rabbit livers, and veal spleen. It also included salted anchovies and capers to balance the bitter taste of the organ meats. These ingredients were then ground up with a meat grinder, resulting in a mixture chunkier than a true pâté.*

Traditional crostini were not made with baguette slices, but with slices of grilled Tuscan bread. First the bread was softened with a small amount of chicken or veal broth. The pâté was then arranged on top and often garnished with one caper at the center of each slice.

Serves 8 as finger food

4 tablespoons (60 g) butter

½ cup (125 ml) extra-virgin olive oil

½ medium red onion, chopped

1¼ pounds (600 g) chicken livers

1 pound (450 g) ground pork

1 (16-ounce) (473 ml) jar giardinera (pickled vegetables)

2 sprigs rosemary, stems removed from leaves

8 fresh sage leaves

1 teaspoon (7 g) sea salt

2 pinches freshly ground black pepper

2 baguette loaves, sliced ½ inch (1.5 cm) thick and toasted

Combine the butter and olive oil in a heavy-bottomed, 4-quart (4 L) braising pan over medium heat. When the butter is melted, add all ingredients except baguette slices; cook, covered, for about 30 minutes, or until meats are well done and vegetables are soft.

ଔ

Remove from heat and allow to cool, uncovered, for 10 minutes; transfer to work bowl of food processor fitted with steel blade. Process until very smooth.

ଔ

Refrigerate until well chilled. Mound the paté onto a serving platter and serve with the crostini.

Filetto di Maiale al Chianti

Chianti Pork Filet

Serves 6 to 8

2 pounds (900 g) boneless pork tenderloin, trimmed of silverskin

Sea salt and freshly ground black pepper

½ cup (75 g) unbleached all-purpose flour

5 tablespoons (75 g) butter

2 garlic cloves, peeled and minced

1 tablespoon (about 5 g) minced sage

1 tablespoon (about 5 g) rosemary

1 teaspoon (5 g) fennel seed

⅓ cup (80 ml) extra-virgin olive oil

¾ cup (180 ml) Chianti wine

Season the tenderloin all over with salt and pepper, then dredge in the flour, coating well and shaking off excess flour. Set aside.

ᘗ

Melt the butter in a heavy-bottomed, deep-sided 5-quart (5 L) sauté pan over medium heat. When the foam subsides, add the meat and cook, turning often, until brown on all sides, about 10 minutes. Add the garlic, sage, rosemary, fennel, and olive oil. Cook for 5 minutes, then add the wine, scraping up the browned bits from the bottom of the pan.

ᘗ

Cook for an additional 10 minutes, or until the wine is almost evaporated and a glaze has formed on the bottom of the pan.

ᘗ

Remove from heat and allow the pork to rest for about 5 minutes before slicing. To serve, slice ½ inch (1.5 cm) thick and drizzle the slices with the pan sauce.

Pandoro di Natale

Pandoro Christmas Tree

There are many stories about the origin of pandoro. *According to some accounts, this traditional Christmas cake comes from Verona, where a type of sweet bread was served by the nobility during the Renaissance; it was called pane de oro ("bread of gold") because the bread was covered by thin sheet of gold leaf. Pandoro as we know it today, a sweet yeast bread baked in a conical, eight-pointed-star mold and covered with powdered sugar, was commercially produced as early as 1894, when the Melegatti company obtained a patent for the process.* Pandoro *can be purchased during the holiday season at many specialty markets.*

Serves 8

2 cups (500 ml) milk

¼ teaspoon (1 ml) vanilla extract

2 eggs

½ cup (100 g) plus 1½ teaspoons (6 g) granulated sugar

¼ cup (35 g) all-purpose flour

2 cups (460 ml) whipping cream

1 plain pandoro

1 tablespoon (8 g) powdered sugar

Candles, candies, Christmas decorations

Combine the milk and vanilla in a heavy-bottomed, 2-quart (2 L) saucepan over medium heat. Heat until milk is warm, then reduce heat to low to keep the milk warm. Combine the eggs and ½ cup (100 g) of the granulated sugar in bowl of electric mixer fitted with wire whisk beater. Beat at medium-high speed until mixture is fluffy and pale lemon yellow in color, about 5 minutes. Add the flour and beat just to blend. Slowly add the egg mixture to the warm milk, stirring constantly with a wooden spoon. Cook over low heat, stirring constantly, until mixture is the consistency of custard, about 5 minutes. Set aside to cool until lukewarm.

೮೨

Whip the whipping cream with the remaining 1½ teaspoons (6 g) granulated sugar. When the cream is whipped, fold into the cold egg mixture to make Chantilly cream.

೮೨

Slice the pandoro horizontally into 5 slices. Place the largest slice on a serving plate to form the bottom layer; spread it with the custard. Cover with the next-largest slice, rotating the layer about 2 inches (5 cm) to the right so the tips of the stacked layers don't line up. Spread with custard. Repeat with each remaining layer, adding the custard between each layer. The result should look like a Christmas tree.

ↅ

Sprinkle the "tree" with powdered sugar and garnish with little candles, candies, and other decorations. To serve, slice from top to bottom; for smaller portions, divide the slices by cutting across them at their midpoint while they are still on the "tree." Transfer the portions to serving plates.

Staff members, family, and friends
(from top to bottom and left to right):
Leonardo. Lavinia, Sarah, Gloria, and
Danila. Tiziana, Pier, Vittoria, and
Francesco.
Tosca. Marco. Johnnie. Gloria.
Alessia, Barbara, and Natalia. Cinzia.
Gena.
Luciano Giannini, Marco. Members
of Amici del Chianti: Gino i' conta-
dino, Miro, Fedora, and Cecco.

About Sarah

When I graduated from high school, my plan was to work on the farm full-time and study at the university in my free time. At that point I knew very little about wine. Of course, I'd harvested grapes with my family as a child, and when we were done they would lift us kids into the cart to crush the grapes with our feet (something that isn't done anymore). Once I began working on the farm, I felt as if I needed to study in order to increase my knowledge about wine in general, so I enrolled in the three-year sommelier course offered each winter by the Associazione Italiana Sommelier (Italian Sommelier Association, or AIS). The classes were taught in a castle not far from my house. At the end of each semester, there was a final exam that you had to pass in order to graduate to the next level. I started out thinking I would just do one year, but I became so passionate about the world of wine that I have never stopped studying it. I discovered that it would be impossible ever to know everything there is to know about wine, to taste all the wines there are to taste and to visit all the wineries there are to visit, and I loved ranking the various producers and comparing notes with other sommeliers. I completed each year of the course, and when I had taken my last exam, I stopped to think about how fast those three years had gone.

I and the other sommeliers had worked hard and learned much. There were eighty of us enrolled at the start, and only forty-two of us got our certificates. In 2010 our AIS delegate, Luigi Pizzolato, recognized my ten years as a sommelier in the Val d'Elsa AIS, calling me "our beloved Sarah, an old young sommelier!" He was referring to the fact that I had been one of the youngest people to take the course, so everyone called me "the young sommelier." But, unbelievably, ten years had gone by since then.

While I was studying to become a sommelier, I also took an eighty-hour course organized by the province of Siena for educational farms. This course covered preparing farms and hiring personnel to make them suitable for student groups of all ages to visit. Upon completion, we weren't ready to start up right away, but our farm did meet all the legal and health requirements by then, so I began drawing up educational materials and using my own imagination to make games and signs. Once everything was prepared, I began sending schools information about our farm and what we could do. Since then, we've had groups of students aged two to eighteen visit the farm.

Class visits to the farm are wonderful. Our goal is to bring children as close to nature as possible and to help them grow into conscientious adults in the future. We give a simple explanation of what organic farming is and have the children participate in all kinds of farm activities. They plant seeds and feed the animals. All the activities are organized along educational paths, and each path has a different theme. It's all designed to explain to the children the concept of a "short supply chain," meaning that we take a product from the farm and arrive at a finished product by doing everything in the same place and not processing things too much. These paths include going from bees to honey, from grapes to wine, from olives to olive oil and from wheat to pasta. There are many others as well.

Meanwhile, after finishing high school I had also started studying at the university in Siena. I enrolled intending to concentrate on foreign languages and literature, but after a year I realized that this wasn't a great choice, because it was preparing me for a future as a professor or as a translator, which wasn't what I wanted to do. I'm not someone who gives up easily, though, so I spent the next two years looking for a course of university study that would satisfy me. Eventually I found the perfect fit: tourism and the economy.

Work on the farm never stopped, and while I had been working on my projects since the year 2000, we began to receive groups of tourists from all over the world who were interested in tasting our wines and visiting the farm.

In 2006, thanks to my friendship with Johnnie and Bruce Weber of San Antonio, Texas, I began to work as a cooking instructor. In 2001 the Webers read about Poggio Alloro in a Fodor's guide to Tuscany and then came for a visit, living with our family for two weeks and taking part in the grape harvest. They quickly became my "adopted parents," and over the years Johnnie has worked with me in the kitchen, in the office, with tours of the farm, and at wine tastings. She constantly promotes the farm as if it were her own. Thanks to Bruce and Johnnie, I began teaching cooking classes and making frequent trips to the United States for this purpose, as well as to promote the farm through wine tastings and other events. Once I'd figured out how to structure and organize cooking classes, I brought the knowledge I'd acquired in the United States back to the farm, and I began to organize cooking classes there as well.

Some funny things have happened in my cooking classes in the United States, especially in the early days when my English was not so great. I often tried to memorize a word five minutes before teaching a class, and when I went to use the word, I would mangle the pronunciation. Once I meant to say that my father is the butcher in our family, and instead I said, "My father is a great bachelor at the farm."

After several years of being encouraged to run for the council of the Associazione Strada del Vino Vernaccia di San Gimignano, I decided to put my name on the ballot in 2009. As if I didn't already have enough to do! Candidates toss their hats into the ring, and then the elections take place immediately. No sooner had I agreed to run than I was elected president. The council consists of a dozen or so members, and one of our main activities is planning the association's major annual event, the anniversary of the San Gimignano Vernaccia wine route. For this year's event, the tenth anniversary, I wanted to show the city's residents, the municipal government, and

tourists what we could do. All of us who were members of the council divided up the tasks involved and rolled up our sleeves.

Frankly, I didn't have any idea how much work my new position would entail, and then organizing this event gave me even more to do. But taking one day at a time and holding steady through some very tough moments, I managed to keep work on the farm going during the peak season and also organize a month of extraordinary events. My days were full, between printing brochures and flyers, drawing up lists of participants and invited guests, collecting funds from sponsors, ordering t-shirts and drinking glasses with the association's logo on them, arranging for advertising, participating in radio interviews, and generally overseeing all the activities and the groups from San Gimignano that were participating.

Running a large association requires a lot of energy, preparation, and the support and faith of other winemakers. I've been very lucky that I haven't been discriminated against because of my age or gender; indeed, I've found people so willing to support me that it's motivated me to work even harder.

Sometimes I don't even realize how many things I do . . . maybe because I'm too busy doing them!

Acknowledgments

This book would not have been possible without the support and love of so many people!

My most heartfelt thanks go to my parents, who put up with me during late-night interviews that lasted for hours, even when their eyelids sometimes began to flutter they were so tired. This book is a special gift from me to you. I never say it out loud to you, so I'm going to write it here: I love you.

Special thanks to Johnnie and Bruce Weber, without whom I would never have had the self-confidence to do all of this. You have always treated me like a daughter and are always telling me how special I am and how much you love me. Thanks also for the wonderful photographs, and for your enthusiasm. You have a special place in the hearts of all of us on the farm.

To my publisher, Kathy Shearer, a woman of incredible patience. You really are an angel. Kathy welcomed me into her home and gave me time to write these pages, and those were some of the happiest days of my life. I also want to express my appreciation to Natalie Danford, my translator, and to Alison Tartt and Barbara Jezek, members of Kathy's production team. Thanks also to Mary Martini and to Johnnie for their help with testing the recipes.

Thanks to my sister Tiziana and my cousin Marco, who have always let me know how proud they are of me.

To my friends Donatella, Majla, Matteo, Marco, Giuseppe, Paolo, Imma, Rossana, Alessio, Sara, Laura, and Filippo (that's chronological order, not order of importance), who have impatiently awaited this book.

To Oriano Stefan, Emiko Davies, Dario Fusar, Leonardo Brogi, Renzo Renzi, and Maggie Shearer for their patience and for giving me their beautiful photographs.

Thanks to the members of my family who offered to tell me more wonderful stories, even those I didn't have time to interview.

Thanks to my cousin Renzo, who works side by side with me, for being patient and shouldering some of my work while I finished writing this book!

Thanks to both old and new friends who have always been there for me and have believed deeply in me.

Bibliography

Balestracci, Duccio. *Breve storia di San Gimignano* [A Brief History of San Gimignano]. Pacini Editore, 2007.

Cavalieri di Santa Fina website: http://www.cavalieridisantafina.it/

Landi, Renzo, and Brunello Bertelli. *Zafferano un fiore che rinasce* [Saffron Blossoms Again]. Nencini, 2001.

Macchi, Carlo, and Bruno Bruchi. *Un Anno con la Vernaccia di San Gimignano* [A Year with Vernaccia di San Gimignano]. Agra, 2011.

Macchi, Carlo, and Bruno Bruchi. *Un anno nel Chianti Classico* [A Year in Chianti Classico]. Agra, 2008.

Mignolli, Luciano. *Il Farro e le sue ricette* [Farro and Cooking with Farro]. Maria Pacini Fazzi, 1991.

Ricci, Nanni, and Diego Soracco. *Extravergine. Manuale per conoscere l'olio d'oliva.* [Extra-Virgin: A Guide to Olive Oil]. Slow Food, 2009.

Sbandieratori dei Borghi e Sestieri website: http://www.sbandieratori.it/templates/www.sbandieratori.it/

Squartini, Bruno. *Città e società: Pensieri sull'anima di San Gimignano* [City and Society: Thoughts on the Soul of San Gimignano]. 2007.

Tocci, Augusto et al. *Tacuinum medioevale. Itinerario gastronomico nella storia* [Medieval Tacuinum: A History of Gastronomy]. Ali&No, 2003.

Tuliani, Maurizio. *La tavola imbandita. Note su alimentazione e società a Siena nel Medioevo* [Setting the Table: Notes on Food and Society in Siena in the Middle Ages]. Betti, 2002.

Vernaccia di San Gimignano Consortium website: http://www.vernaccia.it/

Index

Photography Credits

ALICOLOR, 23

LEONARDO BROGI, 31 (upper left, lower right), 34 (upper right).

CAVALIEIR DI SANTA FINA, 37.

EMIKO DAVIES, 42, 44, 45, 47–49, 51 (upper, lower left), 52 (upper left, right), 54–55, 57 (row 1, left, center right; row 4, center left), 61, 62, 66–67, 77 (row 2, left), 78, 83, 84–85, 88, 89 (lower right), 95 (row 3, center right), 97 (lower), 99 (lower left), 100, 105 (lower), 109, 110 (row 1, right; row 2, left), 113 (row 1, right), 119 (row 4, right), 125, 127, 128 (lower), 137, 139 (lower center), 145 (upper), 147 (left), 170, 173, 174 (lower), 180 (row 2, left; row 3, center), 182 (upper), 185, 186 (lower right), 188, 194 (lower left), 206 (lower right), 207, 217 (row 1, center left, right; row 2, right; row 3, center left; row 4, right), 223 (upper).

FAMILY PHOTOS, 17, 18, 58, 76 (center), 77 (upper left), 142 (lower), 158.

SARAH FIORONI, 40–41, 43 (border photos), 52 (lower), 54–55, 57 (row 3, far right; row 4, center right), 77 (lower right), 79, 86 (right), 89 (row 1, left and right; row 2; row 3, left and center), 90, 96 (upper), 100 (lower), 104, 105 (top, center), 106, 110 (row 1, center left, center; row 3, center), 128 (upper), 130–131, 135 (row 2, left, center left; row 3, lower right), 138 (lower left), 150–151, 153 (row 2, left; row 3, center right), 180 (row 1, center; row 3, left; lower right), 182 (upper), 186 (lower left), 212–213, 217 (row 1, left, center right; row 2, left, center; row 3, left, center right, right; row 4, center), 228 (row 1; row 2, left, center right, right; row 3; row 4, left, center).

DARIO FUSAR 152 (row 4, left), 155.

MAGGIE SHEARER, 34 (row 1, left; row 2, left and center; row 3, left and lower right), 57 (row 2, center left and right; row 3, left; row 4, far left), 86 (left), 103, 113 (row 3, left), 119 (row 1, center right; row 2, left, lower left), 120 (lower right), 124 (upper and lower right), 138–139 (upper), 221 (upper left).

ORIANO STEFAN, half title, title page, 6–7, 9, 10, 24, 27, 28, 38, 39, 43 (center, right), 46, 65 (lower right), 69, 70, 72–73, 75 (row 1, center left and right; row 2, left, center left, far right; row 3, center right and right, row 4, far right), 76 (top, bottom), 77 (row 1, right; row 2, right, lower left), 81, 82, 86 (top, lower center), 95 (row 1, far left; row 2; row 4), 99 (lower left), 102–103, 107, 110 (row 1, left; row 2, center; row 3, right; row 4, right), 113 (row 1, left and center; row 2, center left and right, far right; row 3, right; row 4, left and center left), 114 (top and lower left), 116–117, 119 (row 1, left, right), 124 (upper and lower left), 132–133, 135 (row 1; row 3, lower left), 139 (lower right), 140, 145 (lower), 147 (right), 149, 153 (row 1, left, center; row 2, center left and right; row 3, left and center left, right; row 4, center and right), 154, 157, 158–159, 160, 161 (row 1, left, center left, right; row 2, left; row 3, center right, right; row 4, left), 162, 163, 164, 165, 168–169, 174 (upper), 176–177, 182 (lower), 183, 184, 186 (upper), 190, 192–193, 194–195 (upper, lower right), 197–199, 201 (row 1, left and right; row 2; row 3), 202, 205, 206 (lower left, upper right), 209, 211, 214, 221 (lower), 222, 223 (lower), 227, 228 (row 2, center left; row 4, left), 240.

BRUCE WEBER, 31 (center left, upper right), 51 (lower left), 57 (row 1, center left, far right; row 2, left, row 4, far left), 65 (top), 75 (row 3, left; row 4, center right), 95 (row 1, center right, far right), 99 (top), 135 (row 2, center right), 178, 179, 180 (row 1, left; row 2, center; row 4, lower left), 189

JOHNNIE WEBER, 31 (lower left, center left), 57 (row 2, right; row 3, center left), 65 (lower left), 74 (row 4, center right), 75 (row 1, left), 110 (row 1, center right, lower left), 112, 113 (row 2, left), 117 (inset), 119 (row 1, left; row 2, right; row 3, right), 136, 142 (upper), 147 (right), 153 (row 1, right), 161 (row 2, right; row 4, center right and right), 180 (row 1, right), 201 (row 1, center), 221 (upper right), 231 (upper right).